IMAGINING APC

Also by David Seed

THE FICTIONAL LABYRINTHS OF THOMAS PYNCHON

THE FICTION OF JOSEPH HELLER

JAMES JOYCE'S *A PORTRAIT OF THE ARTIST*

RUDOLPH WURLITZER, AMERICAN NOVELIST AND SCREENWRITER

Imagining Apocalypse

Studies in Cultural Crisis

Edited by

David Seed
Reader, Department of English
University of Liverpool

First published in Great Britain 2000 by
MACMILLAN PRESS LTD
Houndmills, Basingstoke, Hampshire RG21 6XS and London
Companies and representatives throughout the world

A catalogue record for this book is available from the British Library.

First published in the United States of America 2000 by
ST. MARTIN'S PRESS, INC.,
Scholarly and Reference Division,
175 Fifth Avenue, New York, N.Y. 10010

Library of Congress Cataloging-in-Publication Data
Imagining apocalypse : studies in cultural crisis / edited by David Seed.
p. cm.
Includes bibliographical references and index.
ISBN 978-1-349-62247-4 ISBN 978-1-137-07657-1 (eBook)
DOI 10.1007/978-1-137-07657-1

1. Science fiction, English—History and criticism. 2. Science fiction, American—History and criticism. 3. Apocalyptic literature—History and criticism. 4. Literature and society—Great Britain. 5. Literature and society—United States. 6. End of the world in literature. 7. Social change in literature. I. Seed, David.
PR830.S35I43 1999
823'.0876209372—dc21 99–14835
CIP

Softcover reprint of the hardcover 1st edition 2000 978-0-312-22279-6

This book is printed on paper suitable for recycling and made from fully managed and sustained forest sources.

10 9 8 7 6 5 4 3 2 1
09 08 07 06 05 04 03 02 01 00

Transferred to Digital Printing 2011

Contents

Notes on Contributors

Marleen Barr is professor at Montclair State University and a pioneer of studies in feminist science fiction. She has published *Alien to Femininity* (1987), *Feminist Fabulation* (1992) and *Lost in Space* (1993). In addition she has edited and credited a number of essay collections including *Women and Utopia* (1983) and *Future Females* (1981). In 1996 she received the Pilgrim Award for outstanding contributions to science fiction criticism.

Stephen R. L. Clark is Professor of Philosophy at Liverpool University. His publications include *God's world and the Great Awakening* (1991) and *How To Think About The Earth* (1993). He has also written on Olaf Stapledon, among other science fiction novelists, and in 1995 published *How To Live Forever,* a series of lectures on immortality approached through the means of science fiction.

Professor I. F. Clarke pioneered the study of predictive and future wars fiction in his *The Tale of the Future* (1961), *The Pattern of Expectation, 1763–2001* (1979) and *Voices Prophesying War* (1966, 1992). He is currently editing a series of early future wars narrative which includes *The Tale of the Next Great War, 1871–1914* (1995) and *The Great War With Germany, 1890–1914* (1997).

Robert Crossley is Professor of English at the University of Massachusetts, Boston and has published extensively on H. G. Wells, Olaf Stapledon and related authors. His biography of Stapledon *Speaking for the Future* was published in 1994 and in 1997 he edited *An Olaf Stapledon Reader.*

Charles Gannon completed a Ph.D. on future wars fiction from Fordham University after holding a Fulbright Scholarship at Liverpool University 1996–7. He researches on narrative representations of warfare, apocalypse, and advanced technology. He is associate editor of the *Journal of Social and Evolutionary Systems* and is Associate Professor at Eckerd College.

Nick Davis is a member of the English Department at Liverpool University and works mainly on narratology and early modern literature. He is the author of the forthcoming *Stories of Chaos.*

Val Gough is a member of the English Department at Liverpool University. She has edited two collections on Charlotte Perkins Gilman: *A Very Different Story* (1998) and *Charlotte Perkins Gilman: Optimistic Reformer* (forthcoming). She has published numerous essays on women writers, gender and language, and science fiction.

Michael Hoey is a Professor of English Language at Liverpool University. His publications include *On the Surface of Discourse* (1983, 1991) and *Patterns of Lexis in Text* (1991), the latter winning the Duke of Edinburgh award for the best book in applied linguistics.

Veronica Hollinger is Associate Professor of Cultural Studies at Trent University, Ontario. She is co-editor of *Science-Fiction Studies* and co-edited the collection of critical essays *Blood Read: The Vampire as Metaphor in Contemporary Culture* (1997).

Edward James is Professor of Medieval History at Reading University. He is editor of *Foundation: The International Review of Science Fiction* and the author of *Science Fiction in The Twentieth Century* (1994) and co-editor of *The Profession of Science Fiction* (1992).

A. Robert Lee is Professor of American Literature at Nihon University Tokyo. He has edited numerous collections of critical essays for Vision Press on Faulkner, autobiography and African-American Literature. His more recent volumes include *Making American/Making American Literature* (with W. M. Verhoefen) (1996), *The Beat Generation Writers*, and *Designs of Blackness* (1998).

Patrick Parrinder is Professor of English at Reading University and is the author of *The Failure of Theory* (1987) and *Authors and Authority* (1991). He is a leading Wells scholar having served as chairman of the H. G. Wells Society. He has published numerous works on Wells, most recently *Shadows of the Future* which won the Eaton Prize for the best book on science fiction for 1995.

David Seed is a member of the English Department of Liverpool University. He has published criticism on a number of American novelists including Thomas Pynchon and Joseph Heller, and is editor of the Liverpool University Press Science Fiction Texts and Studies. His latest book is *American Science Fiction and the Cold War* (1999).

George Slusser is Professor of English at the University of California Riverside and is director of the Eaton Collection of science fiction. In addition to publishing monographs on Heinlein, LeGuin and others, he has edited science fiction by Conan Doyle, and co-edited numerous critical collections on science fiction including *Immortal Engines* (1996).

1

Introduction: Aspects of Apocalypse

DAVID SEED

'What does the Apocalypse matter, unless in so far as it gives us imaginative release into another vital world? After all, what meaning *has* the Apocalypse? For the ordinary reader, not much.'[1] D. H. Lawrence's questions pave the way for an argument to demonstrate that the Apocalypse *does* matter because it gives us access to a near-defunct symbolistic mode of thought whose rediscovery can re-energize the individual's relation to the cosmos. In his own slim volume *Apocalypse* (1931) he engages in a process of excavation to gain access to the ancient pagan work he is convinced lies embedded within the biblical text building up to a rhapsodic climax celebrating connectedness: 'I am part of the sun as my eye is part of me'.[2] Of course, Lawrence is here pursuing a strategy common to other Modernists of rediscovering (and idealizing) aspects of ancient culture in order to expose absences in the present. More generally, he sets a twentieth-century keynote in interpreting Apocalypse to suit his own preconceptions, and by so doing approaches an oxymoron which will recur throughout this collection: 'secular apocalypse'. Paul Alkon has argued that Jean-Baptiste Cousin de Grainville's *Le Dernier Homme* (1805) secularized apocalypse by altering the relation between ideas and narrative in his work so that, for instance, resurrection was figured as a gradual process, not an instantaneous one.[3] For Alkon the crucial change is an increased narratization of apocalypse, but for the moderns speculative reinterpretation is even more crucial. Lawrence's 1931 work could be read as a guide to those willing to take the trouble to look afresh at the Bible. Similarly Bram Stoker's *The Jewel of Seven Stars* (1903) draws extensively on apocalyptic symbolism to dramatize a revelation of how the religion of Ancient Egypt might live on in Edwardian England. The speculative core of that novel lies in a chapter (Powers – Old and New) that was cut out of the 1912 reprint, which uses recent discoveries like radioactivity to undermine a crudely materialistic worldview, and which speculates on the possible similarities between the state of knowledge at the turn of the century and in Ancient Egypt. Where Lawrence rejects science in favour of a cosmic vitalism, Stoker breaks down presumed distinctions between matter and force, ruminating that even astrology might turn out to be a science after all.

Both these writers juxtapose at least two radically different ways of viewing the world and it may be that one feature of the transition into the modern age was a pluralizing of paradigms. H. G. Wells, such a crucial figure in defining the modern age, as Patrick Parrinder shows here, combines different models to speculate on the fate of mankind in his early scientific romances: Christian, Promethean and evolutionary. The result is a hybridity in his narrative commentary, one which can be found in other science fiction stories from the turn of the century. Simon Newcomb's 'The End of the World' (1903), for instance, describes the imminent collision between Earth and another planet. The protagonist, a professor, and a small group of associates are saved by being in a subterranean chamber when the other planet passes near enough to destroy the surface of the Earth but not the whole planet. Emerging to contemplate the smoking ruins of civilization, the professor glosses the event as inevitable ('such is the course of evolution') and then rationalizes it by reference to a being who is God in all but name: 'to the Power which directs and controls the whole process the ages of humanity are but as days, and it will await in sublime patience the evolution of a new earth and a new order of animated nature'.[4] The human observers did not possess access to this guiding power except in the most general sense and their ignorance of one direction reflected in their miscalculation of the other.

Newcomb's story plays on the two senses of 'end' which inform modern representations of apocalypse: terminus (ending) and telos or ultimate aim. Since the latter remains inscrutable in our most secular fiction, terminus tends to stand in for telos. In *The Sense of an Ending* (1967) Frank Kermode identifies what he calls a 'pattern of anxiety', a recurring perception that we are living at the end of an era.[5] Leslie Fiedler entitled his survey of modern American fiction *Waiting for the End* (1964), reflecting not so much any pre-apocalyptic temper in his chosen writers as his own loss of confidence in an identifiable readership he could any longer address. So he concludes: 'I am inclined to believe that the history of the genre is approaching its end'.[6] In this volume Veronica Hollinger also argues that the science fiction genre appears to have reached its end, but attributes this to generic mixing and the loss of transitivity in some contemporary novels. Reports of the novel's death in the mid-1960s, however, were grossly exaggerated and Ronald Sukenick took appropriately ironic action by converting the proposition into fiction, calling one of his early collections of short stories *The Death of the Novel*.

The notion of an ending, as Kermode shows, does have the appeal of rescuing us from the ultimate nightmare of endless, undifferentiated duration. One tradition which grew from this issue, predating the present age by at least a century, was the series of 'last man' fictions, where the end of the human species is imagined. This was a possibility that appalled H. G. Wells, who concluded his 1893 article 'On Extinction' by praising Thomas Hood's grim poem 'The Last Man' because it evokes the 'most terrible thing that man can conceive as happening to man: the earth desert through a pestilence, and two men, and then one man, looking extinction in the face'.[7] Wells's own time traveller goes many steps beyond, arriving at the penultimate stage of all animal life. He finds himself on a beach peering through the deepening twilight at a black object 'flopping' about on a sandbank. At this point Wells symbolically conflates a number of simultaneous endings: the close of the evolutionary cycle and the light of the sun which is gradually being put out. The time traveller has at least the option of returning to his present to report on his experiences which, significantly, take place on a beach, on the margin beyond which lies the matrix of animal life. W. Warren Wager has described his location as the 'favourite zone for secular eschatologists, marking the point of transition from land to ocean, from man's active life as an air-breather to amniotic unconsciousness and oblivion'.[8] In both Nevil Shute's *On the Beach* (1957) and Helen Clarkson's neglected classic *The Last Day* (1959) nuclear war transforms the beach from a playground to the site for the extinction of life. Shute, who uses lines from Eliot's 'The Hollow Men' as an anticlimactic epigraph, describes the slow but irresistible erosion by fallout of all human habitation. As a character watches the last boat sail towards the greying horizon she reflects: 'This was the end of it, the very, very end.'[9] The most sustained exploration of the beach as a symbolic last point must surely be J. G. Ballard's 'The Terminal Beach', where his protagonist Traven is reminded of his own childhood, wartime bombing sorties, and the more recent use of the island for nuclear tests. The island paradoxically becomes a 'fossil of time future', a place not only of the past, but one where Traven's subjectivity collapses with all dimensions of time, so that his experience, like the text of the story, progresses only in quantal leaps.[10]

If apocalypse underpins the modern sense of crisis, it is clear that sudden or cataclysmic events can be perceived as ruptures in the order of history. I. F. Clarke has shown that the idea of the ultimate war, the war to end all wars, has a history of its own which predates H. G. Wells's popularizing phrase.[11] The Great War – by his own account the most disruptive event in Lawrence's life – was initially named as a single event and then turned into a series which extended as far as World War III, given flesh in a number of works including Sir John Hackett's future documentary duo *The Third World War: A Future History* (1978) and *The Third World War: The Untold Story* (1982).[12] In my essay on Hiroshima I try to show how that event was described paradoxically as simultaneously bringing death and marking the 'birth' of the atomic age. The anxieties about destruction on a previously unimaginable scale proved to predominate in the fiction of nuclear war which, as Paul Brians has shown, gave a fresh impetus to apocalypse, seen in the popularity of 'doomsday' titles and references to ultimate endings.[13] These described the drama of living through the penultimate stage of crisis, represented iconically in the clock face on the cover of the *Bulletin of the Atomic Scientists*, which showed only a few minutes to go to midnight, that is, to zero hour.

The representation of endings in this fiction is ambiguous and never ultimate. For instance, in his survey of the apocalyptic tradition in American Literature David Ketterer stresses that 'apocalyptic literature is concerned with the creation of other worlds which exist, on the literal level, in a credible relationship (whether on the basis of rational extrapolation and analogy or of religious belief) with the "real" world, thereby causing a metaphorical destruction of that "real" world in the reader's head'.[14] And Frank Kermode makes a similar point when he declares that 'Yeats is certainly an apocalyptic poet, but he does not take it literally'.[15] Thus in 'Nineteen Hundred and Nineteen' Yeats evokes an 'end to all things', characteristically turning the notion to include even his own half-written poems in such a way that he avoids privileging his poetry in isolation from the surrounding civil war in Ireland. Both Ketterer and Kermode point out that apocalypse can be read as a figurative account which relates ambiguously to actuality and where destruction might be partial. In contrast to the more obvious reaction of pathos in endings, W. Warren Wagar argues that eschatological fictions help us cope with the fear of death and compensate us for our powerlessness. 'The last man', he continues, 'or one of a handful of last men, is a figure of immeasurable power and importance.'[16] We might also argue that the last man is never a single entity representationally. Even if the last man is his own narrator, he is trying to the very end with the other to which his words are addressed, even if he can pin no specific identity on to that audience. Mary Shelley's Verney imagines being watched by the 'ever-open eye of the Supreme', i.e. the other placed beyond the bounds of mortality; and the narrator of Mordecai Roshwald's *Level 7* (1959) tails off his journal by listing the human companions he has lost: 'Oh friends people mother'.

The notion of apocalypse is closely bound up with our interpretations of time which, as Stephen Clark shows here, might be conceived cyclically, lineally, or through discrete ages. George R. Stewart's 1949 novel *Earth Abides* makes the latter conception unusually explicit. Taking as its premise the virtual extinction of human life by a freak virus, Stewart's novel describes the progress of its protagonist from California eastwards, explicitly reversing the journey along Route 66 described in Steinbeck's *Grapes of Wrath*. Since westward has traditionally symbolized the direction of American settlement, Stewart gives an effect of time rolling backwards with his protagonist; and he regularly punctuates his narrative with interpellated meditation on the non-human time-scales of geological and animal change. Hence the importance of the novel's title, which is taken from Ecclesiastes I.iv, where humans are contrasted with the landscape: 'One generation passeth away, and another generation cometh: but the earth abideth for ever.' The protagonist meets and pairs up with a rare surviving woman and when they discover she is pregnant he is filled with hope for a new beginning: 'Now we have finished with the past... This is the Moment Zero, and we stand between two eras. Now the new life begins. Now we commence the Year One.'[17] Unlike Matthew Arnold, whose visit to the ruined Grande Chartreuse left him 'wandering between two worlds, one dead, the other powerless to be born', Stewart's protagonist relishes his position on the cusp between ages because his perception of the doom of humanity has reversed; 'oh, world without end', he exclaims to himself.[18] Sheer continuity carries a satisfaction itself whereas to

Arnold it simply draws out a limbo state of indefinite duration. Gary K. Wolfe has shown conclusively that novels of post-nuclear destruction usually follow a paradigm of the cataclysm being discovered and surveyed, prior to the settlement and establishment of a new community.[19] Destruction, in brief, is always partial, and the identification of the zero point is regularly a prelude to some kind of renewal.

Destruction too can be read differently depending on its immediate cultural context and the narrative events within which events are situated. Douglas Rushkoff subtitles his *Children of Chaos* (1996) *Surviving the End of the World as we Know it*, a deliberate echo of Cold War rhetoric in order to jolt the reader out of a cosmic fatalism. Stop thinking in terms of entropic endings, he warns his American particularly, and think of the twenty-first century as a new-found land to be explored. Similarly A. Robert Lee traces out the history of the dream of an ultimate war of black insurrection throughout African-American writing as a complex negotiation between spiritual promise and political circumstances. Ralph Ellison's *Invisible Man* (1952), for instance, reaches its climax with a riot in Harlem where a number of apocalyptic parallels expand the meaning of a witness's comment that 'hell broke loose'. The riot clearly constitutes an attack on property and a search for 'loot', but more importantly it dramatizes inversions and disruptions to the narrator's presumptions about reality. So a white figure seems to have been lynched who then turns out to be a tailor's dummy. And all the events take place in near-darkness luridly lit by flames. It appears that the destruction of the old order is taking place, but again the narrator is confused. Near the end of his account he wonders: 'Yes, but what *is* the next phase?'[20] The turmoil of the sixties inner-city riots represented even more clearly to some observers a sign of the Negro Revolution which Martin Luther King expressed figuratively as a cataclysm to the order of Nature: 'the climate, the social climate of American life, erupted into lightning flashes, trembled with thunder and vibrated to the relentless, growing rain of protest come to life through the land.'[21] These sweeping images are placed within a grand narrative of long-awaited revolutionary change.

Holocaust, Martha Bartter tells us, can even play a part in urban renewal.[22] The secularized sequence of destruction of the old order followed by the establishment of a new one is applied constantly in 'aftermath' novels. Here Val Gough demonstrates how Storm

Constantine reinstates aspects of the subculture after a holocaust in order to revise gender roles. In the new prevailing circumstances the prefix 'sub' has to be dropped since holocaust has erased the dominant culture, as happens too in Suzy McKee Charnas' *Walk to the End of the World* (1974) and its sequel *Motherlines* (1978). The former starts with the situation that concludes *Dr. Strangelove*, namely the discussion among top defence officials that they might be able to survive a nuclear holocaust if they go into the deepest shelters. To go into the deepest shelters – and here Strangelove's face would take on a leer – they would need the company of girls with 'sexual characteristics ... of a highly stimulating order'.[23] This is exactly the situation which has already taken place according to Charnas' Prologue. The men hold the power; the women perform their reproductive function. Then, once the holocaust is over, the men take instances from the political events of the 1960s (a black woman starting a civil rights struggle, Vietnamese women fighting with the men, etc.) to articulate a rabid misogyny, displacing their guilt at having produced the nuclear weapons that caused the 'Wasting' onto the women who they keep in servitude to act out the role of scapegoats. Holocaust in Charnas' fiction then pushes women to the extreme of degradation before the sequel can go on to describe a new women-centred society, once they have overthrown the 'rule of order and manly reason'.

Cases like Charnas and Constantine contradict Krishan Kumar's argument put forward in a recent essay where he asks why 'our sense of an ending [is] so flat, so lacking in *élan*.' He continues: 'why have we truncated the apocalyptic vision, so that we see endings without new beginnings?' and concludes that we are living in a period of 'debased millenarianism'.[24] To establish his position Kumar cites Hans Magnus Enzensberger, who sees apocalypse as an entailment of utopia: 'The idea of the apocalypse has accompanied utopian thought since its first beginnings, pursuing it like a shadow, like a reverse side that cannot be left behind: without catastrophe no millennium, without apocalypse no paradise. The idea of the end of the world is simply a negative utopia.'[25] Enzensberger asserts a kind of twinship or mutual dependence between his two concepts, although it should already be clear that the term 'apocalypse' often denotes a complex interplay between endings and beginnings. Not even the most pessimistic apocalyptist closes the door completely to some kind of continuity through rebirth.

Similarly, disaster narratives, J. G. Ballard tells us, have the positive role of describing our confrontation with the 'terrifying void of a patently meaningless universe'.[26] The persistent popularity of tales of planetary collision, nuclear holocaust, and ecological catastrophe would seem to confirm Ballard's assertion. These narratives might take us to the brink of disaster like Chelsea Quin Yarbro's *Time of the Fourth Horseman* (1976) whose title derives from Revelation 6.viii where the fourth seal is opened to reveal the pale horse of death. In Yarbro's novel the analogue with this act is a secret government project to control overpopulation by the covert reintroduction of dormant diseases. Yarbro's novel dramatizes the question whether, as it were, the seal can be closed up again before disaster strikes, whereas in full-blown catastrophe stories 'one can participate in the fantasy of living through one's own death and more, the death of cities, the destruction of humanity itself'. Thus Susan Sontag in her classic essay 'The Imagination of Disaster'. She argues that the appeal of such fantasies lies in the way they address and allay current anxieties about destruction. Her chosen works are specifically science fiction films which, she proposes, use the imagistic potential of the wide screen to produce a 'sensuous elaboration' of fantasy not possible in the printed text. The result is spectacle:

> We may, if we are lucky, be treated to a panorama of melting tanks, flying bodies, crashing walls, awesome craters and fissures in the earth, plummeting spacecraft, colourful deadly rays; and to a symphony of screams, weird electronic signals, the noisiest military hardware going, and the leaden tones of the laconic denizens of alien planets and their subjugated earthlings.[27]

It is rare for these films to deploy biblical parallels as varied as those in Coppola's *Apocalypse Now* where the inversion of images, the portrayal of helicopter gunships as monstrous beings, and other analogues all show the American military adventure in Vietnam to be an exercise in unreason. Usually the spectacle is less intricate. The holocaust of destruction at the end of Bond films routinely celebrates the purging of the evil genius's machinations and symbolizes the restoration of the status quo. And so the examples could be multiplied. Again and again destruction functions as a prelude to restoration.

It also carries the appeal, as Sontag shows, of spectacle and a number of recent narratives have attempted to bring out the ways in which apocalypse can turn into cliché. Anthony Burgess's *The*

End of the World News (1982), for instance, exploits the fiction that the text is an incoherent jumble of genres described by the 'editor' as a 'fantastic tale of the the end of the world, a brief biography of Sigmund Freud' intended for a television series, and the 'libretto of a musical play' about a visit by Trotsky to America.[28] The resultant intercutting between narratives prevents the reader from relaxing into the illusionism of a catastrophe story. Robert Silverberg's ingenious 1972 story 'When We Went to See the End of the World' pursues a different tack. Here the End of the World has been commodified into a time-travel spectacle run by American Express. As friends compare notes about their respective 'trips' there emerges a recurrent mismatch between the signs each couple claims to have seen. Silverberg first implies that the spectacular is a kind of open script on to which each couple projects their favourite disaster. Then he gradually releases snippets of news (city riots, earthquake, a tissue-destroying amoeba at large, etc.) which suggests that some kind of terminal destruction is taking place. But, in the final twist, his characters demonstrate their jaundiced familiarity with the media by dismissing these reports as pure sensationalism: 'you have to expect apocalyptic stuff to attain immense popularity in times like these'.[29] The times are implicitly defined as an era of heightened media consumption that leaves the reader guessing as to the actual state of contemporary America.

The ambiguity of this story reflects the problematic nature of the apocalyptic paradigm. The grand narrative of destruction and renewal seems to function as often as not to challenge readers to identify differences as well as resemblances. Donald L. Moore's novel *Mirrors of the Apocalypse* (1978) is set in a near future where the world is ruled from Jerusalem by an Egyptian-born dictator known as the 'Pharoah' who is perceived to carry some of the symbolism of the perfect being predicted in Revelation. This similarity, however, is countered by his even stronger resemblance with the Old Testament ruler who held the godly in captivity. In so far as he smiles at his enemies (with help of massive orbiting solar mirrors) the Pharoah represents a travesty Old Testament deity; when he tries to abolish religion by fiat he resembles a modern secular dictator. The novel's title plays on these shifting resemblances, tacitly inviting the reader to draw on collective memories of the spiritual history the Pharoah is trying to suppress.

That apocalypse can be both secular and spiritual is shown grotesquely in Madison Smartt Bell's *Waiting for the End of the World*

(1985). Here a motley band of dissidents and psychopaths form themselves into a loose-knit terrorist group centring on Larkin, an epileptic drifter around New York. Larkin both minimizes and inflates his role, claiming the function of a seer: 'I'm just a humble servant of the apocalypse. I've had the visions and seen the signs. Don't you read the papers?'[30] Bell neatly links Larkin's 'vision' to mental disease (he has been an inmate at Belle Vue Hospital) and to the routinization of apocalypse in the popular press. His group steal a number of containers of plutonium isotopes and assemble a crude nuclear bomb in the cellars under New York. Unconsciously acting out the demonized role of Communist saboteurs like those mining American cities in Philip Wylie's *The Smuggled Atom Bomb* (1951), they damage the containers in such a way that New Yorkers start being admitted to hospital with radiation sickness. 'Ruben screwed up the seals,' a member of the gang comments.[31] The colloquialism reverses the trope of opening that usually accompanies apocalypse, revealing nothing. There is no drama of revelation, only a leak. The plutonium literally *is* the gang's power, reified as an object they scarcely know how to handle, and so far from becoming the agents of apocalypse, they fall victim to poisoning or to their own destructive squabbling. Bell signals this process ironically by calling his penultimate section 'Ground Zero' and the concluding part of the novel 'The Ark', the latter gesturing towards a revelation which never comes. The intelligence agencies discover the bomb before it can be used and it is the travesty cult gathered around Larkin which is destroyed, not the hated status quo. Ultimately the actions of Larkin's group can be read as an attempt to reverse their social marginalization by putting flesh on the apocalyptic paradigm ('doomsday in the abstract had little appeal'). Larkin scripts himself into the role of struggling with the devil, masterminds the 'design' of the bombs and even transforms himself symbolically into a part of that bomb. At the very end of the novel Bell still leaves us poised on the border between the spiritual and the secular when he has Larkin die from spontaneous combustion: he 'exploded like a falling star'. Larkin has symbolically become the bomb and is simultaneously linked with Wormwood, the star that falls from the heavens in Revelation.

Bell cleverly assembles a text where there is no authoritative resolution of ambiguity by the narrative voice. Apocalypse becomes part of the text's indeterminacy. Just as Larkin examines the 'signs' of current events, so the reader scrutinizes Bell's narrative to see whether apocalypse can be secularized into pathology or dementia. Robert Crossley shows here how the phrase 'act of God' can circulate through the examination of disasters where survivors speculate about the origins and nature of those disasters. Lois Parkinson Zamora has argued that there is even a near-paradox in that the apocalyptic text gains in significance as it is perceived to be inaccessible. She glosses the act of opening the seals as denoting revelation (the trope of opening up) *and* the closure of meaning by creating a symbolic enigma.[32] Here we return yet again to questions of rhetoric which have been recurring throughout this discussion, and the final aspect of apocalypse that will be considered here is its status as discursive practice.

In *The Sense of an Ending* Frank Kermode explains that 'apocalypse depends on a concord of imaginatively recorded past and imaginatively predicted future, achieved on behalf of us, who remain "in the middest"'.[33] What made his work so useful, as witness its countless citations in the context of apocalypse theory, was its central insight that apocalypse was a narrative, one of the fictions which we employ to make sense of our present. So he argues that 'there is a correlation between the subtlety and variety in our fictions and remoteness and doubtfulness about ends and origins. There is a necessary relation between the fictions by which we order our world and the increasing complexity of what we take to be the "real" history of the world.'[34] 'Doomsday', 'cataclysm', and 'disaster', in short all the routine terms of apocalypse, carry entailments of a narrative without which the words would have no sense, and it is to Kermode's credit that he alerts us to the means by which we articulate this paradigm of destruction and renewal. Lois Parkinson Zamora rightly insists that 'apocalypse' is not synonymous with 'disaster' – although some popular film and fiction make the confusion understandable – but neither does it involve only revelation, despite the term's etymology. Rather 'the apocalyptist assigns to event after event a place in a pattern of historical relationships that ... presses steadily towards culmination'.[35] Promise or prophecy finds its reaffirmation in works like Rene Noorbeergen's *Invitation to a Holocaust* (1981) which 'decodes' Nostradamus to predict that World War III will happen sometime between the early 1980s and 1995, or in Hal Lindsey's earlier bestseller *The Late Great Planet Earth*; although Edward James here demonstrates that Lindsey was drawing on science fiction for his speculative narratives.

The theme which binds all the essays in this collection together, then, is a sceptical awareness of how the apocalyptic paradigm is modified again and again to serve a whole range of interpretive and speculative purposes. Recent study of apocalypse theory is now following a development which can be seen individually in Frank Kermode's revisions of his original discussion. The turning-point by his own account came when he encountered Viktor Shlovsky's notion of 'illusory endings' in the latter's *Theory of Prose*. This alerted him to the way in which endings consist of discursive practices. 'Endings,' he realized, 'are faked, as are all other parts of a narrative structure that impose metaphor on the metonymic sequence.'[36] 'Faking' sounds to carry connotations of bad faith, but is here used more in the sense of 'feign', and it is thus that apocalypse in general has been examined within the context of postmodernism. Richard Dellamora's *Postmodern Apocalypse* (1995) examines different instances of apocalyptic cultural practice and contains among other essays a fine account of how William Burroughs deploys apocalyptic rhetorical strategies in order to uncover homophobia. Jacques Derrida's famous – some would say notorious – essay 'NO APOCALYPSE, NOT NOW' appropriates the language of nuclear strategists in order to show that 'the "reality" of the nuclear age and the fable of nuclear war are perhaps distinct, but they are not two separate things'.[37] Happily, nuclear war has only a potential actuality, and Christopher Norris has recently interpreted Derrida's use of a so-called 'apocalyptic tone' as designed to show the impossibility of moving beyond anthropocentric reality. Specifically, in the essay just quoted Norris sees Derrida's rhetorical moves as recognizing that 'nuclear "reality" is entirely made up of those speech-acts, inventions and projected scenarios which constitute our present knowledge of the future (unthinkable) event'.[38] Apocalyptic texts, then, tend to emerge from such accounts as riven by tensions 'between narrative closure and historical disclosure' (Zamora), as leading towards endings which are ultimately aporias (J. Hillis Miller), or as self-destructive (Peter Schwenger).[39]

Notes

1. Edward D. McDonald (ed.), *Phoenix: The Posthumous Papers of D.H. Lawrence* (London: Heinemann, 1961), p. 294. From a preface to Frederick Carter's *The Dragon of the Apocalypse*.

2. D. H. Lawrence, *Apocalypse* (Harmondsworth: Penguin, 1974), p. 126.
3. Paul K. Alkon, *Origins of Futuristic Fiction* (Athens, Ga.: University of Georgia Press, 1987), pp. 160, 170.
4. Simon Newcomb, 'The End of the World', in David G. Hartwell and L. W. Curredy, eds., *The Battle of the Monsters and Other Stories* (Boston: Gregg Press, 1976), pp. 194, 195.
5. Frank Kermode, *The Sense of an Ending: Studies in the Theory of Fiction* (New York: Oxford University Press, 1967), p. 96.
6. Leslie A. Fiedler, *Waiting for the End: The American Literary Scene from Hemingway to Baldwin* (Harmondsworth: Penguin, 1967), p. 196.
7. Robert Philmus and David Y. Hughes, eds., *Early Writings in Science and Science Fiction by H.G. Wells* (Berkeley and Los Angeles: University of California Press, 1975), p. 172.
8. W. Warren Wagar, *Terminal Visions: The Literature of Last Things* (Bloomington: Indiana University Press, 1982), p. 188.
9. Nevil Shute, *On the Beach* (London: Heinemann, 1957), p. 312.
10. J. G. Ballard, *The Terminal Beach* (Harmondsworth: Penguin, 1966), p. 140.
11. Apart from the essays collected in this volume, see in particular *Voices Prophesying War, 1763–1984* (2nd ed. 1992) and the anthologies *The Tale of the Next Great War, 1871–1914* (1995) and *The Great War with Germany, 1890–1914* (1997).
12. In addition, see Sheldon Bidwell, *World War Three* (1978, cast in the same documentary projection mode as Hackett's volume), and John Stanley's *World War III* (1976), among other works.
13. Paul Brians, *Nuclear Holocausts: Atomic War in Fiction, 1895–1984* (Kent, Oh.: Kent State University Press, 1987), pp. 54–5.
14. David Ketterer, *New Worlds for Old: The Apocalyptic Imagination, Science Fiction, and American Literature* (Garden City, NY: Anchor, 1974), p. 13.
15. Kermode, *Sense of an Ending*, p. 98.
16. Wagar, p. 74.
17. George R. Stewart, *Earth Abides* (London: Corgi, 1973), p. 122.
18. Arnold's image resembles and perhaps dervies from a statement in Thomas Carlyle's 'Characteristics' (1831): 'the Old has passed away: but, alas, the New appears not in its stead; the Time is still in pangs of travail with the New' (quoted in Miriam Allott, ed., *The Poems of Matthew Arnold*, London: Longman, 1979, p. 305).
19. Gary K. Wolfe, 'The Remaking of Zero: Beginning at the End', in Eric S. Rabkin, Martin H. Greenberg, and Joseph D. Olander (eds.), *The End of the World* (Carbondale, Ill.: Southern Illinois University Press, 1983), pp. 1-19.
20. Ralph Ellison, *Invisible Man* (Harmondsworth: Penguin, 1976), p. 464.
21. Martin Luther King, Jr., *Why We Can't Wait* (New York: Signet, 1964), p. 15.
22. Martha A. Bartter, 'Nuclar Holocaust as Urban Renewal', *Science-Fiction Studies*, 13.ii (1986), pp. 148–58.
23. Peter George, *Dr. Strangelove Or, How I learned to Stop Worrying and Love the Bomb* (Oxford: Oxford University Press, 1988), p. 144.

24. Krishan Kumar, 'Apocalypse, Millenium and Utopia Today' in Malcolm Bull, ed., *Apocalypse Theory and the Ends of the World* (Oxford: Blackwell, 1995), p. 212.
25. Hans Magnus Enzensberger, 'Two Notes on the End of the World', *New Left Review*, 110 (July-August, 1978), p. 74.
26. J. G. Ballard, 'Cataclysms and Dooms', in Brian Ash, ed., *The Visual Encyclopedia of Science Fiction* (London: Pan, 1977), p. 130. Brian Stableford argues differently that disasters are always heavily moralized : 'All societies – and perhaps all individuals – sanction the belief that some people deserve to suffer, and that when catastrophe strikes the guilty the moral order of the universe is being conserved' ('Man-Made Catastrophes', in Rabkin, Greenberg and Chander, p. 97).
27. Susan Sontag, *Against Interpretation and Other Essays* (London: André Deutsch, 1987), pp. 212, 213.
28. Anthony Burgess, *The End of the World News: An Entertainment* (Harmondsworth: Penguin, 1983), p. viii.
29. Robert Silverberg, 'When We Went to See the End of the World', in Terry Carr, ed., *Universe 2* (London: Dennis Dobson, 1972), p. 50.
30. Madison Smartt Bell, *Waiting for the End of the World* (London: Sphere, 1986), p. 174.
31. Bell, p. 221.
32. Lois Parkinson Zamora, *Writing the Apocalypse: Historical Vision in Contemporary U.S. and Latin American Fiction* (Cambridge: Cambridge University Press, 1989), pp. 15–16.
33. Kermode, *Sense of an Ending*, p. 8.
34. Kermode, *Sense of an Ending*, p. 67.
35. Zamora, p. 13.
36. Kermode, 'Sensing Endings', *Nineteenth-Century Fiction*, 33.i (1978), p. 147. Shklovsky's main comments on the construction can be found in chapters 5 and 6 of *The Theory of Prose*, i.e. in those dealing with Conan Doyle and Dickens. Kermode has further developed his thinking about endings in 'Waiting for the End' in Malcolm Bull, ed., *Apocalypse Theory and the Ends of the World* (Oxford: Blackwell, 1995), pp. 250–63.
37. Jacques Derrida, 'NO APOCALYPSE, NOT NOW' (full speed ahead, seven missiles, seven missives)', *Diacritics*, 14.ii (1984), p. 23.
38. Christopher Norris, 'Versions of Apocalypse: Kant, Derrida, Foucault', in Bull, pp. 241–2.
39. Zamora, p.15; J. Hillis Miller, 'The Problematic of Ending in Narrative', *Nineteenth-Century Fiction*, 33.i (1978), pp. 3–7; Peter Schwenger, '*Agrippa*, or, The Apocalyptic Book', in Richard Dellamora, ed., *Postmodern Apocalypse: Theory and Cultural Practice at the End* (Philadelphia: University of Pennsylvania Press, 1995), pp. 277–83.

2

The Tales of the Last Days, 1805–3794

I. F. CLARKE

Thirty-six years ago cinema audiences throughout the world saw their worst fears realized in the brilliant, unsparing images of *Dr Strangelove; or, How I Learned to Stop Worrying and Love the Bomb.* The final frames in the Kubrick film displayed the mushroom cloud that signalled the end for all living things; and the last sound from that worst of all possible futures was the voice of Vera Lynn, as she sang of her hope that the world might have a second chance: 'We'll meet again, don't know where, don't know when.'

The up-to-date technology, which underpinned the feasibility of the coming cataclysm in *Dr Strangelove*, concealed the fact that the script of the Kubrick film had recycled one of the most ancient stories known to the peoples of Earth. Long before the many different accounts of the last days of humankind were committed to writing, the tribal lays had recorded the great floods and the other catastrophes that would lead to the End of the World. The Indian version, which figures in the *Mahabharata* and the *Puranas*, expected that the Fire of the Cosmic Conflagration would bring on the dissolution of the universe. This was comparable to the *Ragnarök* myth of the Scandinavians and the *Götterdämmerung* of the Germans – that fearful time foretold in the *Völuspá*, when the last days would bring in 'Fear, malice, and foulness, Ere the World ends. Ere the Doom fall':

> Earth sinks in the sea
> the Sun grows dark,
> the shining stars
> from the sky roll down;
> Steam clouds form
> and fostering fire;
> the flame high rising
> licks heaven itself. [1]

The Greeks refined that simple prefiguration into the Stoic doctrine of the Eternal Return – the perpetual cycle of destruction and renewal. 'According to this theory,' St Augustine wrote, 'just as Plato, for example, taught his disciples at Athens in the fourth century... so in innumerable centuries of the past, separated by immensely wide and yet finite intervals, the same city, the same school, the same disciples have appeared time after time, and are to reappear time after time in innumerable centuries in the future.'[2] Divine intervention in human history, however, had revealed that time would have an end. The dead would rise again. There would be a Last Judgment, and life everlasting thereafter. Time had become linear and all human life had acquired infinite purpose. The Christian account of past and future is a story repeated in countless sermons, many poems, in spectacular paintings and – most striking of all – in the Sistine Chapel frescoes. There the compelling, primordial images of Michelangelo still reveal the beginning and the end of the long journey from the initial *fiat lux* to the final entry into *vitam venturi saeculi.* That late statement from an age of faith was completed some two centuries before Turgot made the first major statement on the idea of progress: 'The whole human race, through alternate periods of rest and unrest, of weal and woe, goes on advancing, although at a slow pace, towards greater perfection'.[3] These notions of progress and perfection were a positive encouragement to go forward in time – to contemplate the more desirable shapes of things-to-come. And so, during the last thirty years of the eighteenth century, a new literature of anticipation began to emerge in the first rudimentary tales of the future. The most popular of these – with four editions in the first year – was Sebastien Mercier's *L'An deux mille quatre cent quarante* of 1771. France in the Year 2440 is no Baconian elsewhere; for Mercier transported his readers from the eternal repose of the old-style static utopia to a world-to-come where no one would ever need to sing *Dies Irae*.

L'An 2440 had immense influence, because it presented a vision of human destiny – a future-perfect society that exemplified the epigraph from Leibnitz: 'The Present is big with the Future'. Mercier creates a revised version of the Eden story. The point of origin in his projection is the central, formative doctrine of the Enlightenment – that the constant augmentation of knowledge is the great engine of human progress. Mercier writes with zest and certainty, as he takes over the controls of space and time – the whole world surveyed and realigned from China to the Americas; all things directed towards the achievement of a final beatitude in time to come. Within twenty years of Mercier's self-assured message from the future, a succession of European writers had followed his example in their visions of the technological advances and social improvements in store for humankind.[4]

The fallout from the first explosion of great expectations obscured the last unsolved question about the future: Would the world go on for ever? The French futurologues had an answer, as they then had answers for every conceivable form of future society. The world would end in accordance with the laws of nature, not by a word from God. So it was written according to Jean-Baptiste François-Xavier Cousin de Grainville in his account of the Last Days of humanity in *Le Dernier Homme* (1805). The circumstances of the writer and the time of writing suggested thoughts of a world in catastrophe. Cousin de Grainville was sixty years of age, a priest who had married to save himself from the guillotine; he had a miserable living as a teacher, and was so desperate about the future that he drowned himself in the Somme canal at Amiens on 1 February 1805.

Cousin de Grainville began writing in 1798, the year of Jenner's *Inquiry into the Causes and Effect of the Variolae Vaccinae* and of the notorious Malthus essay on *The Principle of Population*. The propositions of these two innovators opened up new lines into the future, even to revealing that there could be a limit to human existence; for Cousin de Grainville had perceived how 'the duration of human life was wisely regulated by the omniscient mind of the Almighty, according to the size of the globe and the fecundity of its inhabitants'. His drama of the divine and the scientific gained great power from the immense scale of the action – from Adam to the last human couple – and his narrative of the coming end to human history set the final limit for the imagination in the new literature of coming things. God had foreseen Malthus. His reason for setting a term to life on Earth lay:

> ... in the profound improvements making in medical science by which the lives of thousands of the infantine world have been snatched from the empire of death. and who, in thus becoming the heads of numerous progenies, are laying the foundation of an immense population which the earth in after-ages will be inadequate to sustain. (p. 88)

That expectation became the reality of soil exhaustion and deforestation in the last days of planet Earth:

> The inhabitants of the ancient world, after having exhausted their soil, inundated America like torrents, cut down forest coeval with creation, cultivated the mountains to their summits, and even exhausted that happy soil. They then descended to the shores of the ocean where fishing, that last resource of man, promised them an easy and abundant supply of sustenance. Hence, from Mexico to Paraguay, these shores of the Atlantic Ocean and South Seas are lined with cities inhabited by the last remains of the human race. (p. 141).[5]

As the first future-watchers began to try out appropriate modes for presenting the hopes and fears of the Western world in a time of rapid transition, the new-found practice of conjecture established a gradient of possibilities. These ranged from the might-be in accounts of a French invasion of England to the not-impossible in the first futuristic utopias. At the far-out end there was the planetary journey and – most conjectural of all – the tale of Last Things. That ultimate fiction took well over a century to develop its full potential, since it could not have any connection with human ingenuity until the the atomic bomb made it possible to put an end to all living things. In the first decades of the nineteenth century, however, the prodigious conjecture of *Le Dernier Homme* made for good reading in Europe. There were several imitations of the Grainville story in France; and – in veneration of his achievement – the great French historian, Jules Michelet, devoted a ten-page appendix in his *Histoire du XIXe siècle* to a synopsis of the story.[6] Across the Channel they took up the theme with enthusiasm: Mary Shelley produced her account of *The Last Man. By the Author of Frankenstein* (3 vols, 1826); and there was a remarkable vogue for poems about The Last Man. Thomas Campbell, George Darley, Thomas Hood, Thomas Ouseley, Edward Wallace, and Byron all contributed their visions of the Last Days. 'I saw a vision in my sleep' was the opening line of Thomas Campbell's version of 'The Last Man'; and he went on:

> I saw the last of human mould
> That shall Creation's death behold,
> As Adam saw her prime.

> The Sun's eye had a sickly glare,
> The Earth with age was wan,
> The skeletons of nations were
> Around that lonely man!
> Some had expired in fight, – the brands
> Still rusted in their bony hands;
> In plague and famine some!
> Earth's cities had no sound nor tread;
> And ships were drifting with the dead
> To shores where all was dumb!

A fading sun, a desolate world, empty cities, plague and famine everywhere – these borrowings from Grainville's story became regular entries in the new fiction of Last Things. The stupendous scale of the final drama has always presented a Miltonic challenge of 'things unattempted yet'; and the prophets of apocalypse have never failed to respond by drawing on a prodigious range of possibilities. Their histories of the Last Man have evolved through two main phases in parallel with the changes brought on by the principal regulators of life on our planet: urbanisation, population, industrialisation, technological warfare. Two eloquent images give the visual measure of this movement from the picturesque and eventful to the terrifying and exemplary – from the inevitable disasters of blameless nature to the most guilty consequences of human inventiveness. In the beginning – before ironclad warships and breech-loading artillery – John Martin found a most apt subject for his colossal canvases in his painting of *The Last Man* – a solitary figure, waiting for the end in the midst of universal desolation. In our time, in the most recent phase of these final cataclysms, the cinema has gone forward from Hiroshima to the last days of our world in Stanley Kramer's brilliant film *On the Beach* (1957) – to a solitary submarine tracking along the west coast of the United States in search of the last human beings.

In the first phase of this fiction – from Waterloo to the Somme – the end came naturally and without any help from human ingenuity. Some writers used the theme for the discharge of powerful feelings, but most saw it as an opportunity for excursions into the ultimate and the unknown. The desire for extinction and an end to life, for instance, was central to *Le Dernier Homme*; and in her own narrating of *The Last Man*, Mary Shelley used the limitless opportunities of the future to renew the happiness of her past. In the republican

Britain of the twenty-first century Shelley and Byron are recalled to life as Adrian, Earl of Windsor, and Lord Raymond. As the plague spreads from Constantinople to Europe, they live through the last days of mankind until the narrator, as the sole survivor, is left to end his story with the words: 'Thus around the shores of deserted earth, while the sun is high, and the moon waxes or wanes, the spirits of the dead and the ever-open eye of the Supreme will behold the tiny bark, freighted with Verney – the LAST MAN.'

After Mary Shelley the tale of last things went from one natural disaster to another. A comet ends all life on Earth in Edgar Allan Poe's 'Conversation of Eiros and Charmian'. Camille Flammarion, a most successful populariser of astronomy, has a long account of the future in *La Fin du monde* (1893-94) and that story closes with a remarkable account of the end of planet Earth. Even more striking: the brilliant closing scenes in *The Time Machine* (1895) present the heat-death of our planetary system, as the Wellsian narrator describes how he went forward more than thirty million years into the future, 'drawn on by the mystery of earth's fate, watching with a strange fascination the sun grow larger and duller in the westward sky, and the life of the old earth ebb away'.[7] The red-hot dome of the failing sun, the increased size of the planets, the remote and awful twilight of a dying world – these perturbations in nature come direct from the entry on 'Entropy' in the late nineteenth-century book of science. The precise observations and careful language of the Time Traveller belong to a separate universe, far removed from the hectic, exclamatory narrative of *The Purple Cloud* (1901). There M. P. Shiel tells one of the most extraordinary End-of-Life stories, as Adam Jeffson makes his solitary journey round the world after poison gas from a volcano in the Pacific has killed off all living things. The frantic behaviour of this Last Man – setting fire to great cities is therapy for him – was Shiel's way of giving readers the immediate experience of loneliness. The last Adam, 'afraid of the silence of Europe', continues his search of the world and finds his Eve in Istanbul – a second chance for the last members of the human race.

The most spectacular End-of-the World story, however, was the battle between the secular and the sacred in *The Lord of the World* (1907). The author, Monsignore Robert Hugh Benson, deserves several entries in any book of records: for writing the only account in which God dissolves the world in the twinkling of an eye; for composing an eschatological fiction in which Felsenburgh, the President of Europe, directs the triumphant forces of Anti-Christ; for choosing an Englishman to be the last Pope. On the last day of human history, close to the Mount of the Transfiguration, Pope Silvester says the last mass in the last minutes before the world vanishes.

By 1907 time was running out for the old-style conjectural fiction of the Last Day. In three different and most remarkable stories H. G. Wells gave notice of the immense, ever-growing destructive capabilities of the new technologies. In *The War of the Worlds* (1898) the Martian warriors descend on open sites west of London, armed with the dream weapons of total military power. But Wells stayed his hand: the microbes saved the world. In his second warning tale he took technological weaponry as far as he was willing to go in *The War in the Air* (1908). The German air fleets destroy New York; the whole fabric of world civilization begins to fall apart; famine and the Purple Death wipe out entire populations; the world is brought close to an end – 'Everywhere there are ruins and unburied dead, and shrunken, yellow-faced survivors in a mortal apathy'. The last warning was the far-sighted projection of an atomic war in *The World Set Free* (1914); and again Wells takes his tale of terror to the point when 'most of the capital cities of the world were burning: millions of people had already perished, and over great areas government was at an end'.[8] Wilful optimism then takes over, and out of the chaos of the great catastrophe Wells creates the far better order of the World Republic.

The immense destructiveness of technological weaponry, foreseen in these Wellsian stories, became apparent to all during the First World War. One of the many reactions to that conflict was the sudden shotgun wedding between the fiction of future-warfare and the tales of an end to civilization or to all life on earth.The enormous casualty lists (close on 9 million combatants killed), the entirely unexpected social and political consequences, and the certainty that there would be newer and better weapons – these led to a rapid re-think about the relations between science and society. The heart of darkness was found to lie in the new-found human capacity for creating the most genocidal instruments conceivable. That realization immediately transformed the tale of the Last Days into a most admonitory form of fiction that centres on the dangerous pursuit of super-weapons; and ever since the 1920s, often in most original and compelling ways, writers everywhere have continued to cite *Homo sapiens* as the great enemy of human survival.

The first revised versions of the Last Days were not long in coming. In 1920 Edward Shanks reduced the world to the barbarism described in *The People of the Ruins* (1920). The title tells all: in the broken world of 2074 a series of wars has destroyed most of Europe and reduced London to 'a little settlement containing all that was left of the civilisation of the British Empire'. In the following year, at the National Theatre in Prague, Karel Čapek staged a famous homily on the perils of unreflecting inventiveness. In *R.U.R* the robots represent the final achievement of technology in the false Eden of the progressive, self-justifying society of the future. A rapid fall and final expulsion follow when the robots finish off the human race. 'The end of human history, the end of civilization,' says the General Manager of Rossum's Universal Robots. 'The power of man has fallen,' says the chief robot; 'By gaining possession of the factory we have become masters of everything. The period of mankind has passed away. A new world has arisen.'[9]

As ye have sown, so shall ye reap! The severe morality of this inter-war fiction has spoken plainly in the titles of so many stories – *Ragnarok, Unthinkable, The Black Death, The Collapse of Homo Sapiens, Day of Wrath, The Gas War of 1940, Theodore Savage, Der Pestkrieg, Der Bazillenkrieg, La Guerre microbienne.*[10] These stories of Day Zero are a collective admission of wrongdoing and of false directions. They begin with the expected catastrophes – bombing raids, gas attacks, or secret weapons – and, according to the severity of their judgement, they end with either a pitiful, despairing remnant of humanity or with a small cooperative community, even a new Adam and Eve who are the promise of a better future. Alfred Noyes, for instance, used this lightning transition from the bad old time to the new in order to protest against the immorality of the arms race. 'The result of this was', he wrote, 'that all the combatants, in all parts of the world, possessed a secret weapon so formidable that, to do them justice, most of them would have shrunk with horror from using it, except – and this was the fatal reservation – *except in the last resort*.'[11]

After Hiroshima and Nagasaki that proposition became the key text for the greatest outpouring of warning stories in the history of this apocalyptic fiction.[12] The seemingly imminent peril of sudden extinction united writers throughout the world in telling the one tale of the Last Day. There would be no sound, nothing so archaic as the *Tuba mirum spargens sonum*, nothing more than a flash in the sky. The fire storm follows, and then the mushroom cloud that signals the sudden end for entire nations or – in the worst scenario – for all living things. For more than forty years – from Huxley's *Ape and Essence* in 1949 to the Treaty on Conventional Armed Forces in 1990 – these stories of the last desperate days of humanity raised a permanent question mark over the shape of tomorrow's world. The answers have appeared in many classic accounts of what happened before or after Day Zero. At their best all these stories presented highly realistic, often most uncomfortable anticipations that drew on the abundant literature of nuclear warfare for information about the unthinkable. The Death of Cities, the Devastation, The Time of Fire, World War Terminus – these were some of the names for that time in the near future when civilization would come to a dead end. Then readers would feel the full terror of the Last Day or the anguish of the Day After. The doom-criers saw to it that their stories conformed to the rigours of their intentions: the future offered no hope for feckless humanity or it offered the limited hope of a second chance.

Thus, Aldous Huxley chose to locate *Ape and Essence* (1949) in the year 2108, in the deserts of Southern California, where the degraded survivors worship Belial, the Lord of the Flies. Progress and Nationalism, Huxley says, were the original sins of the modern world; for there can be no doubt that 'the fact of technological progress provides people with the instruments of ever more indiscriminate destruction, while the myth of political and moral progress serves as the excuse for using those means to the very limit'.[13]

There are, however, no survivors in *Level Seven* (1959), since Mordecai Roshwald chose to present his admonitory tale as a day by day account of the last days of life for the obedient button-pushers in the command bunker. Time comes to an end for the human termites after an exchange of missiles. 'I could summarise this war,' the diary reads for June 10, the greatest day in human history, 'in a few words: "Yesterday, in a little under three hours, life on vast patches of the earth was annihilated."' One notable version of life in the United States after The Destruction appeared in *The Long Tomorrow* (1955) by Leigh Brackett. In the small, rigidly controlled communities of the future only the grandparents can recall 'the old days of automobiles, television, aeroplanes, and the bombs that made clouds just like a mushroom'. There is a lesson to be learnt – the hard way. In a calculated imitation of the westward movement of the American pioneers, the young cousins Len and Esau Coulter set off into the wasted areas beyond the Platte River in search of the fabled city of Bartorstown, said to have survived from the time of the Old People. They discover a subterranean nuclear reactor system where the research workers seek a means of controlling 'the interaction of nuclear particles right on their own level, so that no process either of fission or fusion could take place wherever that protecting force-field was in operation'.[14] Bartorstown represents a second chance for the survivors, and that will begin to take effect when all have accepted that self-control is the beginning of survival for individuals and societies.

A most famous variation on this doctrine of human responsibility was the history of challenge and response projected through the history of *A Canticle for Leibowitz* (1959). Walter Miller's narrative advances through the new Dark Ages, from a nuclear catastrophe to a renaissance of knowledge and the fateful rediscovery of atomic energy. 'Are we doomed to do it again and again?' Miller puts the question in the mouth of Abbot Zerchi. 'Spain, France, Britain, America – burned into the oblivion of centuries. And again and again and again.' The citizens of the thirty-eighth century prove to be as incapable of learning from experience as their ancestors of the twentieth century. When the atom bombs go off again, Miller gives the human race a half-chance. The last space-ship takes off with the last children from Earth in keeping with the papal instructions in *Quo perigrinatur grex*: 'The last monk, upon entering, paused in the lock. He stood in the open hatchway and took off his sandals. "*Sic transit mundus,*" he murmured, looking back at the glow'.[15]

In this fiction hope has always been a precious commodity doled out grudgingly by these latter-day prophets. One choice is to be a new-model earthling, reshaped by Natural Selection, a sea-creature happily inhabiting the seas of the Galapagos Archipelago. That is the offer in Kurt Vonnegut's *Galapagos* (1985). Another possibility is to endure the primitive life recorded in Russell Hoban's *Riddley Walker* (1980). Will all life misery be? Do Vonnegut and Hoban have the answers? Time will tell.

Notes

1. Ananda K. Coomaswamy, *Völuspá*. Done into English out of the Icelandic Elder Edda (Kandy: Kandy Industrial School, 1905), p. 11.
2. St Augustine, *Concerning the City of God against the Pagans*, Introduction by John O'Meara. New translation by Henry Bettenson (1972) (Harmondsworth: Penguin Books, 1984), p. 488.

3. Ronald L. Meek, *Turgot on Progress, Sociology and Economics* (Cambridge: Cambridge University Press, 1973), p. 41.
4. There is an excellent account, 'The Secularization of Apocalypse', in Paul Alkon, *Origins of Futuristic Fiction* (Athens and London: University of Georgia Press, 1987), pp. 158–91. The most interesting of the first tales of things-to-come were: Anonymous, *Anticipation, or the Voyage of an American to England in the Year 1980* (1781); Restif de la Bretonne, *L'An 2000, ou la régénération* (1790); A. K. Ruh, *Guirlanden um die Urnen der Zukunft* (1800); Pierre-Marc duc de Lévis, *Les Voyages De Kang-Hi* (1810); Julius von Voss, *Ini: ein Roman aus dem 21en Jahrhundert* (1810).
5. Anonymous (Cousin de Grainville), *The Last Man; or Omegarus and Syderia. A Romance in Futurity* (London: R. Dutton, 2 vols, 1806), vol. 1, pp. 88, 141.
6. The synopsis of *Le Dernier Homme* (pp. 458–66) follows on from the laudatory comments in Chapter X on Grainville in: J. Michelet, *Histoire du XIXe siècle jusqu'à Waterloo* (1875). The imitations of Grainville were: Auguste François Baron Creuze de Lesser, *Le Dernier Homme, poèm imité de Grainville* (1831); Elise Gagne, *Omegar, ou le dernier homme, proso-poésie dramatique de la fin des temps en douze chants* (1858); Alexandre Soumet, *La Divine Épopée* (1841).
7. H. G. Wells, *The Time Machine* (Harmondsworth: Penguin Books, 1972), p. 124.
8. H. G. Wells, *The War in the Air* (London: George Bell, 1908), pp. 345–56. H. G. Wells, *The World Set Free* (London: Odhams Press, 1926), p. 156.
9. The Brothers Čapek, *R.U.R* and *The Insect Play* (London: Oxford University Press, 1961), p. 90.
10. Further details for these stories are:
    Kurt Abel-Musgrave, *Der Bazillenkrieg* (Frankfurt, 1922).
    Professeur X, *La Guerre microbienne* (Paris, 1923).
    Peter Anderson Graham, *The Collapse of Homo Sapiens* (Putnam, 1923).
    Shaw Desmond, *Ragnarok* (Duckworth, 1926).
    Cicely Hamilton, *Theodore Savage* (J. Cape, 1928).
    Miles (Stephen Southwold), *The Gas War of 1940* (Scholartis Press, 1931).
    Francis H. Sibson, *Unthinkable* (Methuen, 1933).
    Moray Dalton, *The Black Death. Sampson Low* (1934).
    Joseph O'Neill, *Day of Wrath* (V. Gollancz, 1936).
11. Alfred Noyes, *The Last Man* (London: J. Murray, 1940), p. 23.
12. Far more comprehensive accounts of the nuclear catastrophe fiction will be found in Paul Boyer, *By the Bomb's Early Light. American Thought and Culture at the Dawn of the Atomic Age* (New York: Pantheon, 1985); Paul Brians, *Nuclear Holocausts. Atomic War in Fiction,1895-1984* (Kent, Oh.: Kent State University Press, 1987); I. F. Clarke, *Voices Prophesying War. Future Wars, 1763-3749* (Oxford & New York: Oxford University Press, 1992); David Dowling, *Fictions of Nuclear Disaster* (London: Macmillan Press, 1987); H. Bruce Franklin, *War Stars. The Superweapon and the American Imagination* (London and New York: Oxford University Press, 1988).

13. Aldous Huxley *Ape and Essence* (London: Chatto & Windus, 1949), p. 94.
14. Mordecai Roshwald, *Level Seven* (London: Heinemann, 1959), p. 125. Leigh Brackett (E. L. Hamilton), *The Long Tomorrow* (London: Mayflower, 1962), pp. 16, 184.
15. Walter Miller, *A Canticle for Leibowitz* (London: Weidenfeld & Nicolson, 1960), pp. 255, 319.

3

The End of The Ages

STEPHEN R. L. CLARK

MAKING THE WHEEL OF TIME

It is a truism that times change, that nothing is ever quite the same again. We can also agree, in principle, that most change is vague: things don't happen all of a sudden, and all together. Even the Fall of Babylon took several days – and some have wondered if it has fallen yet. We might reasonably conclude that the only realistic chronology would be one that counted off the chronons from the Big Bang onwards. There would be no breaks between successive numbers, or ranges of numbers, nor any reason to expect that similar events would be repeated every nth chronon.

But why do we so easily ignore the linear continuum I have described? We prefer to mark off days, years, centuries, millenia and even Great, or Platonic, Years, even though we know that 'the day' does not begin (whether at sunrise or at sunset or at midnight) at the same time everywhere, and that even the terrestrial year has no non-arbitrary beginning. Because the earth's axis rotates, like a gyroscope's, the sun will seem to rise, over the course of a Great Year, against a different stellar background at equivalent moments of the solar year, and gradually trace a circle round the sky until it rises again, most famously at the equinoxes and solstices, within the same zodiacal signs as once upon a time it did (presumably in Babylon). No-one who is anyone believes that these astronomical accidents have any real significance, though enough of our European ancestors thought it mattered to generate wild stories about the passing of Heaven's crown from Bull-god to Ram-god to the Christian Fish (and so, in our own day, to Aquarius).[1] The spring equinox takes a little over two thousand years to move through each zodiacal sign: back before the Bull, it was the Twins that ruled, but we have no record of what religious form that took.

Those are changes that make no material difference. If millenial changes were predictable and real, as they are in Brian Aldiss's

Heliconia trilogy (1982, 1983, 1985), or in Hal Clement's *Cycle of Fire* (1957), then living creatures might, perhaps, have had to adapt to them, as they have in fact adapted to 'the alternation of light and darkness', or to our changing distance from the sun.[2] Conversely, if living creatures had preferred to live thus discontinuously, the millenial changes would have been 'abrupt'. Time is continuous: but life carves it into discontinuous periods. The great polarities of light and dark exist because we have adopted different lives for light and darkness, not because there is an abrupt distinction in levels of radiation as the earth revolves beneath the sun, the moon and stars. Other forms of life might not distinguish night and day at all, nor need to change their habits as the nights grew cold. Others might distinguish more completely: witness Olaf Stapledon's 'plant-men', whose daytime life is vegetable or 'mystical' and whose night-time animal or 'active'.[3] Or rely so heavily on some alteration that we hardly notice (the waning of the moon, or the blossoming of blue flowers) as to carve their phenomenal world, their life-world, into radically different shapes.

On the one hand, these are biological adaptations to cope with drastic changes; on the other, it is the biological choices which mark out the changes as being drastic ones. Without life's presence (so it seems) the universe can only be a flat continuum. Life marks out regions, eras, individuals – and tends to rely on their continuing or recurrent being. But since times change, we cannot reasonably count on Spring or Summer always being what they were. As long as the earth revolves at this particular speed, on this particular orbit, with this particular tilt, our days and seasons may be much the same, because we live the same way as before, or cannot recall how things were different once. The seasons seem the same, just as rivers always seem to run in the same channels even though we all know that they can break their bounds, and run where they do as accidentally as raindrops. Even their flooding, so we tell ourselves, is cyclical: they will be the same again, even when they seem most novel and abrupt.

But of course we do far more than count off days and seasons, floodtime and harvest. We have invented months and weeks, decades and centuries and millenia, and believe in them as readily as we believe the Periodic Table, as though all Saturdays, all Aprils, all Nineties carried the same scent. Some of those variations, maybe, were founded on true observations, of an astronomical or agricultural sort. Once upon a time it was right to say 'when first the Pleiads, children of Atlas, arise, begin your harvest; plough, when they quit the skies.'[4] Once upon a time we could identify, from far in advance, the day when the turtles would hatch, or bulls get randy. Living creatures, long before humanity, discovered the advantages of doing one thing at a time, and making a time to do it together. But weeks and decades do not have that force: what creatures other than the human impose a seven-day periodicity upon their actions, or a five-decade?

There is a sociological or historical study here: who invented weeks, decades and centuries, and why? But my problem is slightly different. Even when there were no weeks (since no-one had invented holidays), and even when there were no centuries (for no-one lived that long), our ancestors believed in Ages. Those Ages might not be exact, any more than a day's work or the rainy season: how long they lasted was not open to exact prediction. Even hours, till we invented clockwork, were of varying length: how much more the Hours of Heaven? Nor do all ages end at the same time everywhere – the Stone Age has continued, in some regions, to this very day. Ages defined the limits of a way of life, long before we thought to specify those limits.

Why should families, nations, species have life-cycles? Considered as collections, or successions, of individuals, such groups are simply aggregates: they do not age and die, because they do not live at all. And yet, we all expect them to, or hope they will:

> Cities and Thrones and Powers
> stand in time's eye
> almost as long as flowers
> that daily die.[5]

Some ends and some beginnings may seem 'natural': the conception or the death of individual organisms seem to be well-defined events, whereas the beginning (or the end) of marriages, or jobs, or church memberships, are socially constructed. The distinction is less secure than it seems: what counts as 'an individual organism' may be determined by ceremonials, or ways of speaking. Until there is something that brings particular sequences together as the experiences of a single creature, the boundaries of that creature, and hence its identity, are moot. So it could be argued that the socially constructed limits of larger 'entities' like cities or civilizations are no different in kind from those of entities we think are 'natural'. Being the same city, under many transformations, may be no more conventional than being the same person. Do caterpillars change into, or give birth to, butterflies?[6] When did you begin: fertilization, implantation, formation of the primitive streak, birth, acknowledgement, baptism or initiation? When will you end: retirement, incapacity, senility, brain-death, burial or the day you are forgotten?

Even if we can, in Plato's words, 'carve reality at the joints', it is clear that we are happy to construct *new* boundaries, in space as well as time. Nothing is real to us, perhaps, that does not have such limits. Marriages aren't real unless they have beginnings, and ends (even if the end is only at what seems a natural death). We might in principle agree that there are differences, and so real changes, which are ineradicably vague. In fact we better had agree to this: if all changes were abrupt, then change, by a simple argument, would be impossible. Consider the last moment of Jeff's not being married, and the first moment of his being married: are these the same moment or not? If they are, then there is a moment (that one) at which he is both married and not-married. If they are different, then – since time is continuous – there are moments in between at which he is neither. Boundaries are magical, because irrational – except of course that 'in the real world' there are no such all-or-nothing changes, and nothing is true only 'at a moment'. Reason (considered as the use of clear distinctions) is as much an alien intrusion on the 'real' as life on the unliving world, and we mark – or make – the event by ceremony.

Some marriages, some cities, some civilizations are begun in deliberate ceremonial. A trench is dug around an area of land, and a sacrifice performed within that area. In other cases, there was no ceremony at the time – but later generations, finding themselves within a city that had simply grown around them, make up a founding legend to confirm that city's being. 'Nothing that once did not exist can exist now unless it once began': and so, we tell ourselves a story of its once-beginning, and by implication of its end. After dissolution all that is left are fragments that late-comers cannot quite imagine were once living parts – except that those ruins themselves may then become a part of some new city, throne or power.

On the one hand, we wish to be part of something that will last forever. On the other, nothing is real to us that does not have an end. The contradiction is solved, it seems, by accepting the cyclical view of being that we impose on history. Every real entity must have an end – but it, or something very like it, comes again.

> And as new buds put forth
> to glad new men,
> out of the spent and unconsidered earth
> the Cities rise again.

Catastrophe may mark the end of an age, and may even do so without any warning-sign, but we still prefer to think that ages come and go. 'There are neither beginnings nor endings to the turning of the Wheel of Time' – in the formulaic words of successive volumes of Robert Jordan's (so far unfinished) *Wheel of Time* (1990-). We here-now are no closer to The End than any other age, although particular, local ends may crowd upon us (and so there is no End). Year, and Great Year, are our images of that continual return: 'Hamlet's Mill', as Giorgio de Santillana called it, in his attempted demonstration that innumerable myths and legends really record our ancestors' grasp of astronomical change – or rather of such apparent changes (mentioned earlier) as are visible from Earth.[7] On the one hand, we are all complacently and happily assured that there can be no real changes, no sudden end to all our plans,

> [and] with bold countenance,
> and knowledge small,
> esteem [our] seven days' continuance
> to be perpetual.

The two convictions, that we are near an End, and yet that Things Go On, somewhere, somehow, generate that image of the Wheel of Fortune. Depending where we think we stand upon that Wheel, we can look forward to decline or triumph – but both states last only for a moment. Even when the scale of events is increased, the pattern remains the same: consider the triumphs and disasters that afflict the several species of Olaf Stapledon's humanity in *Last and First Men* (1930), and of all creaturely existence in *Star Maker* (1937). Contemplation of that Wheel served to distance our predecessors from too great involvement in this-worldly affairs, and too great despair.

SPOKES ON THE WHEEL

The image of the wheel – which probably long preceded any actual potter's or carter's wheel[8] – can be regarded as a compromise between two other images of time: the merely declining, and the open-ended. On the first account we can expect that Everything, as well as every thing, decays down to a drab, undifferentiated and unchanging state. Even protons will one day decay. On the second, growth is unlimited and forever.[9] Current physical evidence suggests that the first account is right, and that any hope we have of Being Forever turns on our escape from Here. So powerful is the image of the Wheel, however, that we constantly prefer to think that even the whole cosmic history is only a phase within a larger cycle, that the cosmos will at last contract into a Big Crunch, and begin, somehow, again. We grasp the existence of This Cosmos by conceiving it as a bounded whole, to be contrasted with Other Cosmoi which can – on the analogy of the returning spring – be conceived as a rebirth of the present.

That 'End of the Whole Cosmos' (and the start of something new) rarely engages mainstream fiction, though occasional poets have written of 'that calm Sunday that goes on and on'.[10] Recent science fiction (e.g. by Stephen Baxter, Gregory Benford, Frederik Pohl) has variously elaborated the theme by imagining that natural entities – not, usually, human – from within one cosmic phase might engineer the next.[11] This world (the story tells us) must inevitably have an end: the only way of achieving Life Forever is to transcend This World, and make or find a new, which will not be subject to the limits of our form of life. The cosmic end, however, is so far away as to diminish our own sense of being: only the most sanguine of anthropocentrists finds it likely that it will be beings of human stock, or anything that looks or thinks like humans, who transcend the cosmos. We can imagine doing it by chance, of course: Poul Anderson's story of *Tau Zero* (1967, 1970), or Stephen Baxter's *Ring* (1994), allow recognizably human beings to move into a brand new universe. But this is only a second chance to achieve transcendence, not the true escape.

It is easier to imagine ends and new beginnings on a smaller scale. When we speak about the End of the World, we usually only mean *our* world. Within our world all ends are also beginnings of a fairly familiar kind, and ends (we think) don't happen all together. Life, we say, goes on, and any partial alteration and destruction will soon be restored. We do have history on our side in this belief: however many millions died in European wars or post-war epidemics in this century there is no lack of people now, nor any clear account of what went differently because they died. Even in a total, global war many peoples and places are, effectively, untouched. However many die, new generations swarm upon us all the time. Even if a way of life departed, after (for example) the Edwardian summer,[12] and the world was changed, so what?

> For the end of the world was long ago –
> And all we dwell today
> As children of some second birth,
> Like a strange people left on earth
> After a judgement day.
> For the end of the world was long ago,
> When the ends of the world waxed free,
> When Rome was sunk in a waste of slaves,
> And the sun drowned in the sea ...[13]

If something ended it must be that something else began. H. G. Wells's minor novel, *Star Begotten* (1937), could be read as science fiction: the Martians (or some other extraterrestrials) invade us by infecting us with their own, irreligious, unmetaphorical mentality. Alternatively, it is only Wells's early, optimistic account of how times change, just as Olaf Stapledon's *Last Men in London* (1932) describes the presence in a young English Quaker of a harder, less sentimental spirit purporting to be a time-travelling Neptunian. We define ourselves by imagining a loathly opposite that looks upon us from Outside: 'intellects vast and cool and unsympathetic'.[14] Writing of Ends that are abrupt, rather than the gradual erosion and mutation that – perhaps - will be our real fate, is to perform that ceremonial act which founds a city, even if (on these occasions) we are unfounding it. The closest that we come to gradual ends that still serve that defining purpose may be in writings inspired by Oswald Spengler. Spengler supposed it possible to identify real, organic entities which animated generations of human individuals to act out a common pattern (spring, summer, autumn, winter) in distinctive styles. Every such culture, he suggested, would exhaust its possibilities, and harden into a world empire, with a strong caste system and an unreasoning conviction that it could never end. A. E. Van Vogt, James Blish, Frederik Pohl and many others have imagined that scenario, often in explicitly Spenglerian terms, and held out the same hope as Spengler, that 'new buds put forth'.[15] We recognize what we are (and can have hope that real life goes on) by knowing when our possibilities, the creative possibilities (for example) of Western or Faustian Humanity, have been exhausted. In James Blish's version the image is literal: the Cities rise up and fly. Weirdly, those Cities are overtaken before their stultification by the real Triumph of Time: that giant which all things devours, 'birds, beasts, trees and flowers'. Yet more weirdly, new worlds bud forth from that catastrophe, of some new sort that Blish (forgiveably) leaves undescribed.[16]

In those visions 'our' end is presaged by the young shoots of another world. Arthur C. Clarke's *Childhood's End* (1954) takes this to extremes: our young, the last generation of humanity, are entirely of another kind. This world is transcended utterly: our seed will take its place in Forever, but only by abandoning, forever, what we were. Conversely, we are defined by the limits placed on our continuance. The theme is repeated, with variations, in Greg Bear's *Bloodmusic* (1985): it is possible to transcend the world, but only by surrendering human nature. The commoner thought hangs on to our humanity, while still conceiving an end to our sort of human being. Those ends are most easily conceived under the heading of catastrophe. Once again, we are not satisfied with gradual transformations and replacements. The end of the age must come abruptly if the age is to be defined: we can't just find that we are somewhere new, and never be able to say when we had crossed 'the border'. Catastrophes, in the fifties, were usually conceived as nuclear spasms, leaving nothing living, or nothing that we could consider worth living. Before and after the fifties writers found it easier to imagine plagues - human or agricultural – which stripped away the support of civilization. Sometimes, like Clarke, they have imagined invasion – more often modelled after European invasion of non-European peoples. In all these ways we simultaneously discover what sort of human beings we are here-now, and what human or almost-human possibilities remain, for good or ill. In George Stewart's *Earth Abides* (1950) the survivors of plague create, mostly without violence, a reasonably civil community; in Algis Budrys' *Some Will Not Die* (1961) co-operation is enforced by violence. And the same old history, despite or because of efforts to recall the past, begins again.

Those catastrophic changes, in imagination, take us back to our beginnings. The darker vision – (Algis Budrys's, or Russell Hoban's

Riddley Walker (1980), or Larry Niven and Jerry Pournelle's *Lucifer's Hammer* (1977)) – supposes that small surviving communities can only defend themselves against brigands by learning the arts of war. A brighter prospect imagines that new forms of culture, and of civil peace, take shape in the aftermath. Ursula Le Guin's *Always Coming Home* (1985) sentimentally imagines both: a fashionable contrast between matriarchal peace and patriarchal violence. John Crowley's *Engine Summer* (1979), with greater originality, conceives many novel forms of civil life, all turning (but by various techniques and to quite different ends) on the need to preserve memory lest worse befall. Catastrophe, we feign, strips artifice away, to reveal an underlying strength, brutal or imaginative. Alternatively, it turns out that we make the same mistakes again: our second chance only confirms our character – as it does in Walter Miller's *Canticle for Leibowitz* (1959).

Some catastrophes – not only nuclear ones – are brought about by human folly. Others seem to strike down from outside, but are easily seen as Judgements, whether they are meteor strikes (as *Lucifer's Hammer*), or plagues (as *Earth Abides*, or John Christopher's *The Death of Grass* (1956)), or alien invasion, as John Varley's *Ophiuchi Hotline* (1977) and Stephen Baxter's *Timelike Infinity* (1992). In all cases our present character and culture prove inadequate to the test, and the End can only be evaded by a change of character and culture that is difficult to conceive.

Another kind of abrupt change will prove to have been heading towards us, unperceived, for decades.

> One summer day, while I was walking along the country road on the farm where I was born, a section of the stone wall opposite me, and not more than three or four yards distant, suddenly fell down. Amid the general stillness and immobility about me the effect was quite startling. ... It was the sudden summing up of a half a century or more of atomic changes in the material of the wall. A grain or two of sand yielded to the pressure of long years, and gravity did the rest.[17]

After that abrupt event we can look back and see its beginnings: beforehand we don't notice any change. The most obvious example of such changes, in recent literature, has been imaginable eco-catastrophe. Temperature rises gradually – but suddenly an unseen line is crossed, and radical change happens. The British

Isles are suddenly tropical, as in Peter Hamilton's *Mindstar Rising* (1993), or else (if the Gulf Stream is diverted) arctic. The oceans, suddenly, are irremediably polluted (as in Gregory Benford's *Timescape* (1980)). Even the meteor strike, in its own way, has that prefiguration.

Some of the stories, obviously, are written only to spur us into action to avoid the peril – but this is to assume that the great change is wholly to be feared. Maybe, as in Gene Wolfe's *Book of the New Sun* (1980, 1981, 1983, 1987), catastrophe will prove our only escape from decadence.

Prefigured changes still allow our past, however redescribed, some meaning: even though we did not, then, know what was gestating, we can look back (or imagine our future looking back) to see the truth. Unprefigured changes, the result of radical, and unpredictable incursions, seem to unwrite the past. None of the plans we made, or could have made, will be fulfilled: the Overlords (of *Childhood's End*) swoop down expressly to prevent our premature attempt to leave the Earth, and nullify all efforts but the Epicurean. The aliens of Varley's *Ophiuchi Hotline* wish only to protect the giant cetaceans (creatures closest of earthly animals to their own multidimensional being), and brush humanity aside without concern, and also without malice. Such invasions are imagined, exactly, to deconstruct our ordinary trust that things are, somehow, on our side. Invasions of this sort signify that the world is alien to any human interest, or even to any rational interest at all. In Baxter's *Ring* (and its associated stories) humanity cannot compete successfully with the 'lords of baryonic life', the Xeelee – but the Xeelee too are doomed. In Bear's *Eon* (1985) and *Eternity* (1989) it is the genocidal enemy of humankind that will, in the end, be dominant. Either there is no significance at all in human life, or else 'significance' shrinks down to particular human moments, having no lasting influence on anything beyond. In that moment of disillusion we may believe that 'here we have no continuing city', that our destiny lies outside time entirely. The sheerly nihilistic conclusion may, nowadays, be commoner.

BREAKING THE WHEEL

One Age, in our imagination, succeeds another. In some time to come, things will be different, and everything that we have done or been already will have a different meaning. Our Western civilization began, as Chesterton observed, among the ruins of an older world – and even that older world, the Classical Western World, began among the ruins of an older still. The curtain of history rises on a world already ancient, full of ruined cities and ways of thought worn smooth.[18] Sometimes we imagine that things were really different then, in ways that we can't now imagine or believe.[19] Sometimes we fear that we are indeed repeating history, that our future will lie in ruins, from which our remote descendants, if they are lucky, will remake a world – and see it fall to ruin. The other, apocalyptic, vision suggests that there will be an End: somehow, sometime, the world will be made new in a way that does not lead once again to ruin. 'I saw the new Jerusalem coming down from heaven.'[20] In R. A. Lafferty's *Fourth Mansions* (1969) it is imagined that the Cycle of destruction is the 'failed or broken cycle': success is 'to be taken up to heaven in every stone and person'.[21] In Niven and Pournelle's *The Mote in God's Eye* (1974), their unfortunate aliens, the Moties, are condemned to cycles of destruction because they cannot escape their own biology, nor yet their planetary system. No doubt their creators intended a moral for humanity: vulgarly, that we need to leave the earth; anagogically, that we need to escape the cyclical way of thinking.

Not all new beginnings (even ones that don't repeat the past) offer true transcendence. The end of the Third Age in J. R. R. Tolkien's tale of Middle Earth does not put things back to their beginning: there are true losses, true novelties, and a widening future. The Fourth Age, and its successors, are the Age of Man – that is, of mortal creatures made responsible for the wellbeing of their world. Once upon a time the gods took care for us, as once before they had even walked among us. Once upon a time there were immortal creatures in the woods and mountains: fallible, and fractious, to be sure, but still preserving knowledge of our real situation, here amongst immensities. Now, we are left to get along with things: 'the deeds of Men will outlast [elves and dwarves] ... and yet come to naught in the end but might-have-beens'.[22] A similar thought appears in Michael Scott Rohan's *The Winter of the World* (1986–8), associated with the imagined end of the last Ice Age (which was almost the final Ice Age). Indeed, the thought has appeared in many different forms, and with more or less enthusiasm, in many works of science fiction. 'Drake is no longer in his hammock',[23] the First Ones have departed,[24] and everything is up to us, for good or ill. In much science fiction this is, deliberately, couched in atheistical terms: Blish's *Black Easter* (1968) carries this to extremes by envisaging the victory of the Satanic forces at Armageddon (because 'God is dead'), and then wrapping up the moral, in *The Day after Judgment* (1972) by suggesting that God had wanted Man to take His throne, and hoped to force the outcome by allowing Satan's victory. This thought, however comfortable to human arrogance, is heresy. Tolkien's example demonstrates that similar thoughts are still compatible with orthodox theism.

The same may be true for the dream of transcendence. The stories I have just described offer a radical transformation: the movement from adolescence into adulthood (and the risk of failure). But ordinary living, so it seems, disguises adulthood: few of us are conscious of an absolute responsibility for our conduct in the world, or modify our lives to take account of that. Mostly, we are asleep, and differ from adolescents only in that there is no-one now to get us up for school. The last days are signalled, in R. A. Lafferty's *Apocalypses* (1977), by the fact that 'the more than four billion inhabitants of the world ... simply do not act as if we had had these wars'.[25] The challenge posed by some writers and thinkers is to wake up – and unsurprisingly, few manage to write any plausible account of what it's like to be awake. Gordon Dickson imagines the cusp: the age when (literally) every human individual has to become a god (or all humanity will die). The story prevents its own denouement: not knowing how to be a god, how can we write, or read, of what that's like? The climax is constantly being put back: *The Final Encyclopedia* (1984) fails to show the outcome; *Chantry Guild* (1988) comes closer; but Dickson must then turn aside to examine earlier episodes of his imagined future. If he could describe what will be we would have reached the end already – and it is clear we haven't.

This may be a further reason for the nihilism or defeatism I mentioned before: it can't be human beings like us who go on past the End, and the only way of representing such an End is by imagining the end of us, our being displaced by aliens or artificial forms. In Baxter's *The Time Ships* (his 1995 sequel to Wells's fantasy), even a 50-million-year head start (by stranding a small human colony that long ago) only results in the destruction of Earth's biosphere, and human beings' replacement by their own von Neumann probes, evolved to high intelligence.[26] Either way the beings that we now are must end: either we shall be changed entirely, or we shall be displaced – and who could tell the difference?

Rather than face that prospect we imagine a retreat: a return to the World We Want. The most recent form of imagined alteration is a step into Another World, a world entirely plastic to the human imagination: some versions rely on a wholly fictional technology, others on an extrapolated version of contemporary science. In the first sort of story, the discovery of 'hyperspace' allows us interstellar travel – but also (and more interestingly) a step into a radically non-spatial world, the Other World.[27] In the second sort, the Internet expands into the 'consensual hallucination' of cyberspace, where nothing exists accept what is humanly meaningful (for good or ill). Both sorts of story, in effect, appeal to the world of magic, and promise an end to Endings.

In the 'real' world (as moderns have imagined it) there are no privileged moments, nor any causes but the mechanistic. Our predecessors thought otherwise: for them, the world of time reflected, though imperfectly, eternal forms, and there were moments when those forms were visible. Our minds too reflected the eternal, and were equipped to understand the world because we shared its origin. The conviction that we could understand, that we could find the principles of things 'within ourselves, as Plato taught', as Galileo said, has played a crucial part in the development of science. Without that faith, it seems unlikely that our theories could have much to do with the non-human world. Maybe moderns are right to seek a different, and unhuman, understanding of the world – but the task may be beyond our powers. The older pattern always reasserts itself: the end of geological ages is connected with catastrophe, and those in turn with stories about descending stars. Every 20 million years or so, a dark companion of the sun (named Nemesis) disturbs material in the Oort cloud, and new comets begin their long slow fall toward the inner solar system – coinciding not infrequently with Earth. The Earth is wiped clean, and a new age begins in fire and steam – as the older legends always said was true.

The moral, for our imagination, seems obvious. If we think only about the 'real' (modern) world, and the sort of mind appropriate to it, we are condemned to believe that something utterly unhuman is appropriate. Gregory Benford's mech-wars stories[28] propose a lasting conflict between mechanical intelligence and living purpose. If we prefer to carry on conceiving something like a human world, we imagine a new kind of human: 'new buds' that have recovered hope. Rebirth is a sleep and a forgetting: forgetting that we've reached an end. Either vision is appropriate to science fiction: the genre's frequent failures are when neither happens, and instead new ages and new regions turn out to be peopled by all the old characters, with all the old confusions.

By Santillana's account most tales of gods and heroes – and catastrophes – concern the stars. That is where Real Creatures can be found, and their history too is one of change. The Flood attested by universal legend did not drown our mountains, but the heavenly places: another star takes up residence as the axle-pin; another sign disappears behind the sun as it rises at the spring equinox; other signs dip down below the horizon, and are drowned. Up There the Real Heroes chase their prey around the Mill of God, which is rebuilt once every two millenia or so. In our day (as I said before) the sun is beginning to rise, in spring, in Aquarius; later on Polaris will slip away from its present role as Pole Star; there will be new rulers in the heavens. Of course, we know that this is all appearance, the last gasp of Babylonian astrology, but it still has meaning.

The ancients could assume that Here was central to the cosmos: not that Here was the most important place – far otherwise, it lay down at the very bottom of the world, and real events were played out Overhead. But at least the cosmos, so described, was a familiar one. Looking outward to the stars we saw a world – not ours, but still a human world. Nowadays the world we see Out There is quite unhuman. We take it for granted that nothing There has any living explanation: to suggest that the mass predicted by our theories, but missing from any astronomical observation, might have been enclosed in Dyson spheres (for example) by galactic civilizations, or that any astronomical event is best explained as the act of living beings, is no more than a joke (although we can't say why). Even the Ophiuchi Hotline would, most probably, be reckoned just a 'natural event'.[29]

Virtual Reality (so called) is a retreat from this immensity: a consensual hallucination of the sort admired by Orwell's Party leaders: 'The earth is as old as we are, no older. How could it be older? Nothing exists except through human consciousness.... The stars ... are bits of fire a few kilometres away. We could reach them if we wanted to. Or we could blot them out.'[30] Hyperspace, conceived as the Place behind all places where the true originals of everything enjoy their lives, is devised as a way of reaching other worlds, ones not confined by human consciousness. Unfortunately, it is usually conceived – in literary fiction – that we cannot reach, from Hyperspace, to any world that we cannot conceive.[31] Really alien worlds – that is, really alien ways of feeling and understanding – are excluded. The effect may therefore be the same as Virtual Reality: we hide our eyes from something greater and stranger than we can bear.[32]

Suppose we manage not to retreat, and let the heavens be what they are. Even the small steps we have managed to take Out There have let us look back on the beautiful blue bubble that is Earth and realise our consanguinity. That icon ('Earthrise on the Moon') has probably played a larger part in ending the Cold War, and alerting millions to the risk of ecological catastrophe, than any careful writings. At the moment cosmological speculation about Big Bangs, Black Holes, Naked Singularities and so on do more to fuel human conceit than weigh it down. It is more marvellous that we (or some of us) should be able to uncover the cosmos than humbling that the cosmos is so much larger, and so different from what we thought. But if we really came to believe that There is the real world would we not begin to understand our ancestors?

> When you have left the womb of the lower world, will you not attain to what you formerly conceived as ideal? On this journey you will first encounter the blue which is commonly supposed to be the ceiling of heaven; but this is ignorance. For in reality it consists of the most refined region of the air which appears to be this colour owing to its distance, as experts in optics tell us. When you have passed through this tract you will discover the stars to have been created not merely to sparkle but to be new worlds. And at last, to prolong the account no further, the infinite itself will unfold.[33]

The End of the Age, or of the Ages, will lie in the discovery of Forever: we shall not inhabit that Forever in the forms we now possess. Acclimatising ourselves to live beneath the eyes of heaven will be a spiritual exercise as difficult as any, or else a wished-for transformation. Growing Up, on this account, will not simply be to take full responsibility for 'our own lives' (as in an earlier venture): it will be to recognise how little of 'our lives' are ours, how much the worlds roll on regardless (and ourselves with them). The End of the Age will dawn in the discovery, the deep acknowledgement, that there are now no Ages, that things are forever. By John Crowley's evocative account the moment when Giordano Bruno fully realised that the Sun did not revolve around the Earth was his release from the crystal spheres that bound all human souls. Instead of having to clamber, in imagination, upwards to the heavens, he realised that the Earth itself was swimming *through* the heavens, that he had already escaped. 'You made yourself equal to the stars by knowing your mother Earth was a star as well; you rose up through the spheres not by leaving the earth but by sailing it: by knowing that it sailed.'[34] A similar revelation may help to release us from our fear that the End is Nigh: it has indeed already happened. 'The end of the world was long ago.' The End of the Age is the end of all the ages, for it subverts that ancient cyclicism. For better or worse, 'Crazy Eddie has won his eternal war against the Cycles'.[35] Science fiction has tended to represent that ending in material or atheistical terms, and so to exaggerate the alien nature of whatever sensibility is more appropriate to 'timelike infinity'. But the breakout from our crystal palace has long been anticipated in religious fiction.

> And for us this is the end of all the stories. and we can most truly say that they all lived happily ever after. But for them it was only the beginning of the real story. All their life in this world and all their adventures in Narnia had only been the cover and the title page: now at last they were beginning Chapter One of the Great Story which no one on earth has read: which goes on for ever: in which every chapter is better than the one before.[36]

Which is the reality, and which the metaphor, is a matter on which reasonable folk may disagree.

Notes

1. Another way of putting the same point is to say that the Virgin will have a son – that is, the sun rises in Virgo at the autumn equinox if it rises in the Fishes at the spring (Virgil, *Eclogues* 4). Virgil's poem was taken to be an unconscious prophecy of the birth of Christ.
2. In the former, seasons last for centuries; in the latter two distinct but inter-related ecosystems have evolved to suit temperate and very-high-temperature ages.
3. Olaf Stapledon *Star Maker* (London: Methuen 1937), pp. 153 ff.
4. Hesiod, *Works and Days*, lines 383–4: T. F. Higham's translation.
5. Rudyard Kipling *Verse: 1885–1926* (London: Hodder & Stoughton 1927), p. 479 (the poem is the prelude to *Puck of Pook's Hill*).
6. Naomi Mitchison answers this Aristotelian question by imagining, in *Memoirs of a Spacewoman* (1962), a species of sapient caterpillar that

gives birth to a butterfly which will be immortal, unless the caterpillar managed to have sex before pupation (if it did, the butterfly dies in giving birth to caterpillars).

7. Giorgio di Santillana and Hertha von Dechend *Hamlet's Mill* (Boston: David R. Godine 1977); John Crowley's *Little Big* (1981) transforms some of Santillana's thoughts into high art.
8. Poul Anderson *Trader to the Stars* (1964) imagines a society where the astronomical wheel is sacred, and the idea of using a model wheel for transport is anathema. Maybe that was the Mayas' problem too.
9. The view which C. J. Cherryh attributes to her imagined Unioners: 'generations whose way of life was stars, infinities, unlimited growth, and time which looked to forever' (*Downbelow Station* (London: Methuen 1981), p. 14.
10. J. B. S. Haldane *Possible Worlds* (London: Chatto and Windus 1927), p. 312, uses the passage, from Swinburne, to describe one of his two futures: 'either the human race will prove that its destiny is in eternity and infinity' or else the end will come. Arthur C. Clarke concludes his *Prelude to Space* (1953), with an obscure reference back to Haldane, and votes for Swinburne's conclusion.
11. Stephen Baxter *Ring* (1994); *The Time Ships* (1995); Greg Bear *Eon* (1985); *Eternity* (1989); Frederik Pohl *The Annals of the Heechee* (1987).
12. Though Kipling had long anticipated the 'end of deference' which later writers think is distinctively modern: see 'Chant-Pagan' (*op. cit.*, p. 453), the words of an English veteran of the Boer campaigns.
13. G. K. Chesterton, 'The Ballad of the White Horse': *Collected Poems* (London: Methuen 1950), p. 225.
14. As H. G. Wells describes his Martians in *The War of the Worlds* (Harmondsworth: Penguin 1946, p. 9): what he there describes as enemies, to be defeated, he and Stapledon elsewhere describe as our spiritual superiors: see also Doris Lessing's *Shikasta* (1979).
15. A. E. Van Vogt *The Voyage of the Space Beagle* (1951); James Blish, *They Shall Have Stars* (1957); F. Pohl and J. Williamson, *The Reefs of Space* (1963).
16. James Blish, *The Triumph of Time* or *The Clash of Cymbals* (1958).
17. John Burroughs, cited by Bill McKibben, in *The End of Nature* (Harmondsworth: Penguin 1990), p. 5.
18. A claim I have elaborated in 'Ancient Philosophy': *Oxford Illustrated History of Western Philosophy*, ed. A. Kenny (Oxford: OUP 1994), pp. 1–53.
19. As John Crowley supposes in his *Aegypt* sequence: so far, *Aegypt* (1987) and *Love and Sleep* (1994).
20. Revelation 21: 2.
21. *Fourth Mansions* (New York: Ace Books 1969), p. 102.
22. J. R. R. Tolkien *The Lord of the Rings* (London: Allen & Unwin 1966; 2nd edition), vol. 3, p. 149.
23. Susan Cooper *Silver on the Tree* (1977).
24. *Babylon 5*: see especially 'Into the Fire', in the fourth series. The story echoes the conclusion of E. E. Smith's *Lensmen* series, in which Eddorians and Arisians alike must make way for the younger peoples.

25. R. A. Lafferty, *Apocalypses* (New York: Pinnacle Books 1977), p. 354: Lafferty here conceives that the history of our century was 'at first' only a series of tragi-comic operas created by Enniscorthy Sweeny, under the influence of demons.
26. The pyramidal structure of his Universal Constructors, the inheritors of what was once human, is reminiscent of the similar machines in Pohl and Kornbluth's *Wolfbane* (1959), where they are more usually the villains.
27. See especially Gordon Dickson, *The Final Encyclopedia* (1984), and *Chantry Guild* (1988); Vonda MacIntyre *Superluminal* (1983); Melissa Scott, *Five Twelfths of Heaven* (1985). Robert Heinlein, in *Waldo* (1950), and Isaac Asimov, in *The Gods Themselves* (1972), use hyperspace explicitly to permit access to unlimited power (at an eventual cost) rather than unlimited flight.
28. Beginning with *In the Ocean of Night* (1971), and continuing through *Across the Sea of Suns* (1984), *Great Sky River* (1987), *Tides of Light* (1989), *Furious Gulf* (1994) and indefinitely many others. Like Saberhagen's Berserkers Benford's mechs may someday be redeemed, transformed into discrete and passionate intelligences.
29. See Stanislaw Lem, *His Master's Voice* (1968). The fact that no civilized beings should expect their descendants to inhabit a galactic empire is not itself a reason to deny that there could be one: if there ever is such an 'empire', it probably exists already. 'Where are its emissaries, then?' If they were here, we wouldn't be. When they arrive, we shall be changed.
30. O'Brien speaks: Orwell, *Nineteen Eighty-Four* (Harmondsworth: Penguin 1954), p. 213.
31. Melissa Scott, *Five Twelfths of Heaven* (1985) and its sequels explicitly identifies Hyperspace with the symbolic universe of Hermetic magic: see my *How to Live Forever* (London: Routledge 1995), pp. 122 ff.
32. See Olaf Stapledon, *Last and First Men* (London: Methuen 1930), p. 311 f. on members of the last human species traumatized by an attempted interstellar flight: 'they fill their minds with human conceits, and their houses with toys'.
33. Edward Herbert, *De Veritate*, tr. M. H. Carré (Bristol: Arrowsmith 1937), p. 329.
34. John Crowley, *Aegypt* (London: Gollancz 1987), p. 366.
35. L. Niven and J. Pournelle, *The Mote in God's Eye* (Futura Publications: London 1976; first published 1974), p. 560.
36. C. S. Lewis, *The Last Battle* (Harmondsworth: Penguin 1964), p. 165.

4

Rewriting the Christian Apocalypse as a Science-Fictional Event

EDWARD JAMES

The anonymous author of *The Epic of Gilgamesh*, Homer, Plato, Lucian of Samosata, Jonathan Swift, and many others, have at one time or another been proclaimed as progenitors of science fiction: no-one in print, as far as I know, perhaps out of misplaced reverence, has suggested St John of Patmos, the author of the Book of Revelation.[1] Yet he is one of the most widely quoted and influential of all writers on the future: the symbolic creator of a prophetic tradition that has influenced much more secular approaches to speculation about the future, and his Book survives to this day as an influential and powerful way of imagining the future. Utopian thinkers and activists have drawn on the apocalyptic theme of the millennium for their visions of the perfect world; the dramatic tales of cosmic struggle to be found in the Book of Revelation are comparable in the sweep and the 'sense of wonder' that they evoke to the most extravagent space operas of the science-fictional tradition. In this paper, I shall look at some of the interactions between what we might see as rival eschatologies: the vision of the end of all things presented in Jewish and Christian revelation (the Greek word for which gives us 'apocalypse') and the view of the end of all things presented by science fiction.

There is one unresolved contradiction between the two, which ultimately relates to the scientific context in which these alternative visions originated: the Christian view, even in twentieth-century prophetic literature, views the 'end of all things' almost invariably in terms of the end of the Earth, while the science-fiction viewpoint might well have an eschatological endpoint which encompasses the death of the entire universe, most typically in a contraction of all matter which mirrors the Big Bang origin of that matter. The imagery of revelation in Jewish and Christian texts, necessarily since the texts reflect a pre-Copernican world-view, see the end of the universe in terms of a crisis in human politics and the ultimate cleansing of the Earth by God's will. The stars in John's Revelation do not fall into a singularity at the centre of the universe: 'the stars of heaven fell unto the earth, even as a fig tree casteth her untimely figs, when she is shaken of a mighty wind' (Rev. 6: 13). That geocentric outlook is only treated ironically or satirically by science fiction writers:[2] for scientific reasons, there is a very real and virtually unbridgeable divide between the Biblical and the scientific visions of the 'end of all things'. The only time that I have seen the Book of Revelation viewed in a perspective that took account of the universe beyond the solar system is among the Mormons, where it has been suggested that after the Last Judgement the Earth will become ablaze with light and glory, like the great stars in the heavens, which have already passed on to their exaltation.[3] But Biblical and scientific visions do converge if one looks in science fiction for stories of the death of the earth, or the death of human civilisation, rather than the death of the entire universe, and that is what this chapter is largely concerned with.

St John of Patmos was, of course, by no means the first to write a vision of future history: he came at the end of a long line of apocalyptic writers, whose works have been enshrined in the Hebrew Bible, the Jewish Apocrypha, and the New Testament. There are two texts that have been particularly significant in later Christian eschatological speculation: the Book of Daniel, and Revelation itself. The Book of Daniel, probably written around 164 BCE,[4] contains a number of prophecies relating to the end of the world, which have been almost as influential among more modern prophecy believers than those in Revelation. Daniel 2 tells of a statue with a head of gold, chest and arms of silver, belly and thighs of brass, and legs of iron with feet of clay and iron mixed together: the head of gold symbolises the kingdom of Nebuchadnezzar, says Daniel, and the other parts of the body the three kingdoms which will follow him. The final kingdom will break up, destroyed by the stone 'cut out without hands' (2: 34) and God shall 'set up a kingdom, which shall never be destroyed' (2: 44). Chapter 7 tells of four beasts, the fourth of which has ten horns, including a little horn with a 'mouth that speaketh great things' (7: 8): this beast is destroyed by the 'Ancient of Days'. The four beasts are the four kingdoms; the ten horns are like the ten toes of the statue, the ten kingdoms which will contest for power in the last days. Chapter 9 speaks of the seventy weeks that shall intervene before the arrival of 'the prince that shall come' (9: 26) – the Anointed One (Heb: Messiah; Gk: Christos). The best known prophecy comes in Chapter 11, which tells of the rise and fall of the king of the north. This king, who invades the south and places the 'abomination that maketh desolate' in the sanctuary, will be arrogant and impious, but, 'at the time of the end' (11: 40),

> he shall come to his end, and none shall help him. And at that time shall Michael stand up, the great prince which standeth for the children of thy people: and there shall be a time of trouble, such as never was since there was a nation even to that same time: and at that time thy people shall be delivered, every one that shall be found written in the book. And many of them that sleep in the dust of the earth shall awake, some to everlasting life, and some to shame and contempt. (Daniel 11: 45–12: 2)

Biblical scholars for a long time have seen all these prophecies as interpreting in terms of a great cosmic drama the actual historical events between Alexander the Great (the statue's legs and the ten-horned beast) and Antiochus Epiphanes, who desecrated the Temple and placed 'the abomination that maketh desolate' within it. But for many interpreters, Daniel has been a rich source of imagery by which their own times and the immediate future can be understood.

St John's Revelation is much more explicit than that of Daniel, and much better known, particularly as a result of its frequent representation in the art of the Middle Ages, on the portals and walls of so many cathedrals or in the wonderful illuminated manuscripts of the *Commentary on the Apocalypse* by Beatus of Liebana. At the opening of the vision, the throne of God stands in front of a glassy sea, surrounded by twenty-four elders and four beasts. On the right hand of God rests a book with seven seals. As each seal is broken, new catastrophes are let loose upon earth. The breaking of the seventh seal brings forth the 144,000 humans who were destined to be saved; but the destruction of the earth continues. Eventually Satan brings two Beasts from the sea, one of which rules men for forty-two months, marking everyone on the forehead with his mark ('and his number is Six hundred threescore and six': 13: 18). The final conflict between the kings of the earth ensues, at Armageddon in Palestine, and despite the appearance of the Great Whore, Babylon falls, and 'the Word of God' (Christ) comes on a white horse to defeat the forces of evil. Satan is imprisoned for a thousand years, and Christ reigns with the saints on earth during this Millennium (20: 4). When Satan is released, he leads another battle against God, together with Gog and Magog, and after his defeat the Final Judgement takes place, 'and whosoever was not found written in the book of life was cast into the lake of fire' (20: 15). The new Jerusalem descends from heaven, twelve thousand furlongs long and wide and high, with its jewelled walls and its twelve pearly gates, where the faithful may live for ever and ever (21).

Daniel and John have provided Christians with a richly ambiguous and powerful imagery from which to construct ideas about the future and explanations of the present. Other books of the Christian Bible contribute other ideas: the figure of the Antichrist, for instance, first appears in 1 John 2: 18.[5] But it is noticeable that the theologians of the Early Church became increasingly worried by the emphasis that some Christians gave to these Biblical prophecies. Towards the end of the fourth century the North African theologian Tyconius wrote the first extended commentary of the Book of Revelation. Tyconius was a Donatist, whose writings were systematically destroyed by the Church which persecuted them, but in the fifth century Gennadius of Marseilles said of Tyconius' work that it explained John's book 'in a spiritual sense, without anything carnal in it', and 'removed the idea of a thousand-year future kingdom of the just on earth'.[6] Augustine seems to have followed his fellow North African Tyconius quite closely, first in his Letter to Hesychius, and then in *The City of God*, 20: 7–10. Thereafter, literal ideas of the end of the world and the millennium of Christ's rule on earth seldom found a welcome within orthodoxy. The orthodox theological position today has recently been clearly stated by David Fergusson:

> The traditional scriptural images of the end are problematic if we suppose them to be detailed, albeit coded, messages from the future. The apocalyptic images of the Bible are too opaque, too inconsistent and too conditioned by their original *Sitz-im-Leben* (life-context) to be considered cryptic statements about how the world will end. Christian faith and hope, moreover, are predicated upon our ignorance of exactly how the future will be.[7]

As a result, therefore, according to Fergusson, literalist interpretations of the Apocalypse tend to be found on the edges of orthodoxy:

> Throughout the history of the Church ... there have been fringe groups proffering a more detailed and urgent eschatology usually in criticism of the established church.[8]

Sometimes believers in the prophecy attain influence through their writings, like Joachim of Fiore in the twelfth century; at other times, these beliefs inspire social or political action, as in the various millenarian groups in the Middle Ages, famously discussed by Norman Cohn,[9] or with the various English revolutionaries of the seventeenth century, investigated most notably by Christopher Hill.[10]

It is interesting that Christopher Hill, so sympathetic to the beliefs of those outside the mainstream of society, should nevertheless feel the need to apologise for exploring 'a trivial blind alley in human thought', even though 'at all points it trembles on the edge of major political issues'.[11] In the first edition, he had been even more dismissive of this blind alley, noting that such beliefs eventually 'disappeared ... into the world of cranks';[12] similarly, the medieval historian Marjorie Reeves noted in 1969 that 'prophecy has now ceased to be of importance, except on the fringes of modern civilization'.[13] Both these comments were singled out by Paul Boyer in the prologue to his study of modern prophecy belief in the United States of America, who remarked that in the USA in the late twentieth century 'the "fringes" were very broad indeed, and the "cranks" numbered in the millions!'[14] To a European, the figures quoted by Boyer are startling (although hardly more startling than comparable figures for beliefs in creation 'science' or in alien abduction). In 1986, 29 per cent of whites believed the Bible to be literally true, and 44 per cent of blacks; 48 per cent of southerners described themselves as 'born-again Christians' (as compared to 19 per cent in the north-east); a study of a 'not particularly conservative' community college in Oregon in the mid-1980s found that one-third of the students believed in the Rapture (a peculiarly American addition to eschatological ideology, as we shall see below); overall 50 per cent of American *college graduates* [Boyer's emphasis] awaited Christ's Second Coming.[15]

Given the ubiquity of prophecy belief in American culture, the influence of Biblical catastrophe on science fiction writers, if only in terms of titles, as Warren Wagar has pointed out, is hardly surprising.[16] Noah's flood, later interpreted as a prefiguring of Armageddon and the Last Judgement, for example, features in the titles of Garrett P. Serviss's *The Second Deluge* (1912), Wells's *All Aboard for Ararat* (1941), and William Tenn's 'Generation of Noah' (1951), and the events of Revelation in Stephen Southwold's *The Seventh Bowl* (1930), Joseph O'Neill's *Day of Wrath* (1936), Edmond Hamilton's 'Day of Judgment' (1946), Philip K. Dick and Roger Zelazny's *Deus Irae* (1976) and others. To Wagar's list one might add many others: to take Armageddon alone, there is (in order of publication) Philip Francis Nowlan's *Armageddon 2419 AD* (1928; 1962), Mordecai Roshwald's *A Small Armageddon* (1962), Hilbert Schenk's *A Rose for Armageddon* (1982), Mike Farren's *The Armageddon Crazy* (1989) and David Weber's *The Armageddon Inheritance* (1993), together with at least two anthologies, Michael Sissons, ed., *Asleep in Armageddon* (1962), and Walter M. Miller, Jr., ed., *Beyond Armageddon: Survivors of the Megawar* (1985). The last title, of course, is a reminder that the use in a title of words like Armageddon or Doomsday (which is the English for 'Day of Judgement') is usually not even a token reference to the Book of Revelation, but merely a common shorthand for a man-made catastrophe, particularly a devastating war.

Occasionally, science-fiction writers have borrowed wholesale from Christian mythology: examples would be James Blish's *Black Easter* (1968) and *The Day after Judgement* (1972), in which Hell materialises in Death Valley, California, and Robert A. Heinlein's *Job: A Comedy of Justice* (1984), where the protagonist visits the Heavenly City described in Revelation 21. More recently, James Morrow has been exploring Christian mythology in each successive book, from *This Is the Way the World Ends* (1986) through *Only Begotten Daughter* (1990), *Towing Jehovah* (1994), its sequel *Blameless in Abaddan* (1996), to a series of short stories published as *Bible Stories for Adults* (1996). Each of these authors deal with their subject using the rational or scientific mode of science fiction; but, of course, they are not dealing with the world as it would be recognised by conventional *or* Christian scientists. At this point, therefore, one starts investigating the ever-subjective dividing-line between science fiction and fantasy.

The endless debates on this matter suggest that there can be no agreed answer. A case in point would be Msgr. R. H. Benson's *The Lord of the World* (1907).[17] According to Bleiler, this follows not so much the Biblical prophecies as the text known as *The Prophecies of*

St Malachy (which have nothing in reality to do with that twelfth-century Irish monk).[18] Here advances in technology and progress in science are shown to have clear connections with the deline of religion. Felsenburgh, the Antichrist, becomes ruler of the world, and sends bombers after Pope Sylvester, who is in hiding near Megiddo (Armageddon). The ending is unclear, but apparently God manifests Himself, and the faithful are taken up into heaven.[19] A prophecy believer who is a reader of such science fiction might, in theory, accept this as legitimate and rational extrapolation from known precepts about the world; most science fiction readers would not. This is no doubt why the numerous reworkings of similar ideas in post-war popular fiction, are normally treated as either fantasy or as part of the horror genre, despite their strong affiliations to science fiction: *The Omen* series of books and films (which began in 1976), Stephen King's *The Stand* (1978), and Daniel Easterman's *The Number of the Beast* (1992) are obvious examples.

As stated at the beginning, science, and hence science fiction writers, operate on quite a different cosmological scale from the writers of Biblical apocalypse. 'The end of all things' has nothing to do with the end of the world: the Earth will almost certainly have been consumed by an expanding Sun long before the death of the universe, however that happens. Few writers have depicted the end of the universe. Two that have done it in their novels are James Blish and Poul Anderson, and it is interesting that both of them do use Christian imagery, and both of them do see the catastrophe in terms of a new beginning, just as St John did. Blish's *A Clash of Cymbals*,[20] in the original edition, described the end of the universe as happening in 4004 CE: as the protagonists died in the final collapse, one of them pressed the detonator in his space suit, and the novel ended with the words 'Creation began'.[21] The cyclical element is emphasised by the date, which repeats the date worked out by the seventeenth-century Archbishop Ussher for creation, 4004 BCE.[22] Poul Anderson's *Tau Zero* (1970) features a (Swedish) starship which goes out of control, and is unable to prevent its acceleration to far beyond the speed of light. Thanks to the effects of relativity, its crew is able to witness the death of the universe, and the ship is able to survive into the creation of a new universe, and to begin a new human civilisation. Anderson is skilled enough to avoid too-obvious parallels: the planet is no Eden – and none of the new colonists is called Adam or Eve...[23] But religion is duly recalled in the chapter in which the universe ends, as the crew sing a bawdy song about St Peter at the pearly gates, and the captain's response to the knowledge that there will be a new beginning is 'Father, I thank Thee.'[24]

Biblical apocalypse is much more frequently recalled in science fiction in the stories of catastrophe on earth, when human civilisation is wiped out either by natural causes – by plague, or collision with an asteroid – or by its own actions – some scientific experiment that goes wrong or, most commonly, catastrophic war.[25] Here too the holocaust is frequently followed by a rebirth: holocaust novels are far more frequently novels about post-holocaust survival and perhaps revival. There are, of course, many other reasons for this than the desire to parallel the story of Noah in Genesis 9 or the story of the two episodes of quasi-utopian rebirth in Revelation 20 and 21. It is not easy to write a novel which ends with the total annihilation of humanity, as, for instance, Mordecai Roshwald did in *Level 7* (1959) or Thomas M. Disch in *The Genocides* (1965), nor does this ultimate downbeat ending sit well with the expectations of popular fiction. A post-holocaust novel has the advantage of allowing the author to recreate the world as he or she wishes, rather than to go through the chore of having to extrapolate current trends into a plausible future. Holocaust wipes out the problems of the present, to create a new, possibly simpler and, from the point of view of both author and characters, more manageable world. Characters are immediately put into a situation of conflict and struggle; the characters immediately have the clear goals of survival and the rebuilding of civilisation. The author can work out his or her social and political ideas on a clean slate: it is a way of creating a utopia, or a potential for utopia, without having to worry too much about the process. 'Utopia' should be qualified: most of these are bucolic or Edenic arcadias rather than anything like classic utopias. J. Leslie Mitchell's *Gay Hunter* may stand as an example, where those who survive the holocaust end up basking in an idyllic Chiltern landscape.[26] Such an arcadia corresponds, of course, to some of the Biblical prophecies, which have lain behind many utopian visions over the centuries: visions of rural idylls, where 'the wolf shall dwell with the lamb' (Isaiah 11: 6). There is more than one utopia in the Bible, however, which 'begins with a narrative of Utopia as a garden, [and] ends with a vision of Utopia as a city',[27] and the direction which Utopia takes is ultimately subjective and personal. The 'survivalist' novels of the 1980s, are utopias for some. They are associated with those groups who actually prepared for post-nuclear survival in the backwoods of America, and who in the 1990s became associated with the Militia movement. Here the holocaust is used to cleanse the world of its corrupting forces (liberals, feminists, homosexuals, and blacks) and restore the good old masculine values of the (fictional) American West. As Clute has noted, 'Sadistic, sexist, racist, pornographic, gloating and void, survivalist fiction is an obscene parody of genuine survivalism, and a nightmare at the bottom of the barrel of sf.'[28]

Survivalist fiction only exaggerates to extremes the tendencies of post-holocaust fiction as a sub-genre, for such fiction has its disturbing elements. An aspect of the sublime, which several commentators have seen as one of the essential elements of science fiction, is the pleasure obtained from the contemplation of the infinite or the almost inconceivably large: the death of civilisation would surely fall into that category too. There are also perhaps the even less savoury pleasures in destruction itself, which may have its roots in the urge towards self-oblivion, and a pleasure in the prospect of surviving when others have died. Some of these aspects of holocaust have been explored by science fiction writers. There is the wonderful 1946 story 'Vintage Season' by Catherine L. Moore, for instance, in which time travellers flock back into the past to visit New York, for the pleasure of seeing the nuclear missiles fall on the city (this was reworked by Robert Silverberg as 'In Another Country'),[29] while the normality of holidaying in a catastrophe epoch was well satirised by Robert Silverberg in 'When We Went to See the End of the World'.[30] Well worth noting in this context is 'The Big Flash', a short story by Norman Spinrad.[31] The narrator is an agent for a new rock group, the Four Horsemen. The video of their first major release, *The Big Flash,* intersperses shorts of the group with shots of corpses at Auschwitz and in Vietnam, and ends with a shattering sound and the image of an exploding nuclear bomb: indeed, *all* their videos end like that. The US President realises the opportunity he has to use the group to soften up public opinion on the tactical use of nuclear weapons; a cult grows up around the group, whose symbol is the mushroom cloud and the words 'Do It'. The story ends with the plan to drop a test-bomb at the climax of the Four Horsemen's fourth concert. The captain and men of a Polaris submarine are hypnotised by the televised concert: eventually, as the frenzied audience shout 'DO IT!', the order is given to press the button. The last words of the story are 'THE BIG FLASH'. This story not only expresses the author's feeling that there is a general fascination with self-destruction, but also links this with the religious impulse towards the Endtime. Indeed, there is the hint that the Four Horsemen are what they say they are: if not the Four Horsemen of the Apocalypse themselves, then performing a similar divine or Satanic function.

At first sight there can be little in common between science-fictional ideas of the Endtime and those of Christians who believe in the relevance of Biblical prophecy to their own time. Few science fiction writers have strong religious beliefs, and almost none would think of the Bible as the literal word of God. If they use Biblical images and ideas, they do so because they know that those images are bound to strike a powerful resonance with their American readership. And the prophetic visions in the Bible, and perhaps particularly those in the Book of Revelation, apparently have little to do with the post-Darwinian and post-Einsteinian world-view of the great majority of science-fiction writers.

However, this conclusion would be false, because it would not take account of the direction which American apocalyptic thought has taken over the last century and more.[32] As elsewhere in the Christian world, both pre- and post-Reformation, there was a division between post-millennialists and pre-millennialists. The former believed that Christ would come at the end of a millennium of peace, and that it was the duty of Christians to prepare for His coming by perfecting the world around them: this chimed in well with the nineteenth-century belief in progress and in the manifest destiny of America, but was also a natural doctrine for those Christians who saw their duty in social reform and the betterment of the world. Pre-millennialists believed in a more literal interpretation of John: that before Christ's coming there would be great destruction and death, and that all Christians could do was to reform themselves and to pray. The former strand was very influential in America in the nineteenth century; but it tended to be adhered to by those for whom biblical prophecy was not very important and the Second Coming a very long way in the future. Pre-millennialists were much more inclined to prophesy themselves, and to preach the immediacy of the Endtime. William Miller (1782–1849) was one of the most influential of these; his preachings led directly to the foundation of the Seventh Day Adventists, and indirectly to the origins of the Jehovah's Witnesses. But arguably even more influential, despite the fact that the sect that he founded (the Plymouth Brethren) was a small one, was the Englishman John Darby. His 'dispensationalism' was spread through the United States not only by his own preaching, but by the indefatigable preaching of his followers. Darby taught that none of the Biblical prophecies applied to the current age, or dispensation, but that all those events foretold by John and others would come to pass very quickly as soon as the next dispensation began, with the Rapture. The Rapture – when believers rise up bodily to meet Christ in the air, and thus avoid the final Tribulation – had appeared before as an idea, but had never been given any prominence. Unlike Miller, Darby also brought all Biblical prophecies together into one system; thus he taught that the Jews would re-establish their homeland in Palestine, and begin rebuilding the Temple, and that, after the Tribulation, the surviving remant of Jews would accept Christ as their Messiah.

Pre-millennialists, whether influenced by Darby or not, did not set any store by secular reform or progress, and were inclined to see disorder in society as a sign – almost a welcome sign – that the end was near. The war which began in 1914, and the Balfour Declaration in 1917, were both seized on with fervour. In the 1930s, we meet what may be the first attempt to enrol the novel on the side of preaching. Forest Loman Oilar's *Be Thou Prepared, For Jesus is Coming* (1937) told of how a committee was set up in a typical American city to investigate a rash of disappearances: they discover that the faithful have left the ungodly behind, taken up to Heaven in the Rapture.[33] As in the Book of Revelation, when the Antichrist takes over, he brands people with his Mark: in his novel Oilar adds the detail that those who refuse branding are stripped, sprayed with sulphuric acid, and decapitated. This novel investigates some of the practical consequences of the Rapture, as later novels in the genre do, such as the slow death of pets, locked up in the houses belonging to the saints who had been taken up. It is certainly possible to detect a certain sadistic delight in the torments of those not caught up in the Rapture, as the science fiction satirist Thomas M. Disch has expressed in his poem 'The Rapture':

> Lift us higher make us see
> The vistas of Eternity
> How the godless squirm in hell
> Their meat napalmed into a gel[34]

In a number of Rapture novels, it is the city which collapses first: the rural audience for prophecy preaching will naturally see the city as the natural home of Antichrist.[35] In this investigation of the consequences of 'What if?', novels like Oilar's are just as logically rigorous as the classical science fiction novel, even if (like quite a number of science fiction writers) they do not accept the orthodox scientific world-view: it is interesting that to my knowledge neither their authors nor science-fiction critics treat them as having anything to do with science fiction.

Prophecy preachers incorporated many of the events of the period from the 1930s onwards into their vision of the Endtime: the Depression, Hitler's war, the foundation of the state of Israel, and, above all, the invention and use of the atomic bomb. As II Peter 3: 10 prophesied, 'The heavens shall pass away with a great noise, and the elements shall melt with fervent heat, the earth also and the works that are therein shall be burned up.' In some senses, therefore, above all in the prominence of nuclear holocaust and in the emphasis on dystopic futures in general, Christian and science-fictional prophecy had converged. The optimism of the nineteenth-century post-millennialist preacher or of his secular equivalent, the utopian visionary, all but disappeared. As a prominent pre-millennialist wrote of post-millennialists in 1970, with typical lack of clarity:

> These people rejected much of the Scripture as being literal and believed in the inherent goodness of man. World War I greatly disheartened this group and World War II virtually wiped out this viewpoint. No self-respecting scholar who looks at the world conditions and the accelerating decline of Christian influence today is a 'postmillennialist'.[36]

That quotation comes from Hal Lindsey's *The Late Great Planet Earth* (1970), which we may take as our example of post-War prophetic preaching. It was the best-selling non-fiction book of the 1970s in the USA, with 9 million copies in print by 1978 and 28 million by 1990.[37] Lindsey starts his book by musing on the idea of the future and how it intrigues us. 'In our imagination we long to step out of our humdrum existence and into worlds beyond. Take science fiction as an example. It fascinates us.'[38] The future can only be understood through the Bible, however. Lindsey goes on to show (by denying the validity of the past century of Biblical scholarship) the success of Biblical prophets: he terms them 'Israeli prophets', with an eye to his future exposition of the importance of modern Israel.[39] Most Jews of Jesus's day ignored the clear prophecies of the Israeli prophets, because they were blind, indifferent, and prepared to follow their leaders unquestioningly.[40] Now we have equally clear prophecies, which predict a sequence of events preceding the endtime, including the foundation of the state of Israel, the capture of Old Jerusalem by Israel (which happened in 1967, notes Lindsey), the rebuilding of the Temple, and war with the various powers mentioned in Daniel, interpreted as Russia, the Arab states and China (discussed in a chapter entitled 'The Yellow Peril'). The ten-nation confederacy symbolised by the ten toes of the statue in Daniel's vision is clearly the European Economic Community. In his discussion of the coming end, and on how to recognise the Beast, the Antichrist, Lindsey plays on many fears: the fear of crime and drugs, of over-population, of communist conspiracy (suggested to be behind the ecumenical movement), and of nuclear war. Lindsey's message is that none of this, not even nuclear holocaust, need worry the faithful: they will be raised up in the air to meet Christ in what Lindsey calls 'the ultimate trip'.[41] They will miss the final war, when Russia invades the Middle East ('see chart one for the movement of troops'),[42] but they will return again with Christ for His millennial rule on Earth. Some other prophetic writers were even more optimistic, suggesting that America would be largely spared from nuclear destruction, and that its major problem would be that so many of the leaders of American society were God-fearing people who would be removed from Earth in the Rapture that America would be left very short of leadership.[43]

Lindsey has received a great deal of criticism for his book from more orthodox Christians. 'The blatant disregard for this world, the avoidance of the cost of Christian discipleship and the implicit anti-Semitism have all been noted.'[44] A recent study of the rhetoric of apocalyptic by Stephen O'Leary mercilessly exposes Lindsey's fallacious reasoning and Bibilical criticism.[45] But O'Leary suspects that Lindsey's influence has been considerable, even though this would be difficult to prove. A 1984 survey of American attitudes towards nuclear war found 39 per cent of respondants agreeing with the statement 'When the Bible predicts that the earth will be destroyed by fire, it's telling us about nuclear war.'[46] Few Europeans realised how suffused were Ronald Reagan's speeches with the style of rhetoric adopted by Lindsey and other pre-millennialists, nor how willing he was to interpret current events in apocalyptic terms.[47]

Not only was there a convergence of Biblical prophecy and science-fictional 'prophecy' in the 1970s, in terms of some specific endtime visions, but convergence has continued through the 1980s and 1990s, though in rather different directions. Lindsey himself moved in a new direction at the very beginning of the 1980s, in his book *The 1980s: Countdown to Armageddon.*[48] While continuing to stress the coming of Antichrist (who 'is alive somewhere in Europe; perhaps he is already a member of the EEC parliament'),[49] he also allows in what he calls 'A RAY OF HOPE'. Suddenly it is possible to take an active role, to stem the tide of Biblical prophecy, to 'stop the Soviet's insane rush toward nuclear war'.[50] Evangelist Pat Robertson made the same shift in attitude in the mid-1980s, in his preparation for his presidential bid in 1988: effectively he moved from a premillennialist lack of interest in worldly affairs and secular change to a post-millennialist interest in reform and in giving 'every American a vision of hope'.[51] This shift was a necessary one, perhaps, as a number of prominent prophecy believers moved into the unfamiliar arena of political activism, and had to modify their premillennialist ideology in order to accommodate ideas of reform and progress. The collapse of the Soviet Union, which for decades had played the major role in endtime scenarios, also had a role in changing the tone of prophecy preaching. The apocalyptic message is still there, but the urgency and immediacy has to some extent evaporated, as well as the clarity of the vision.

It may be stretching the comparison, but to some extent this has been true within science fiction in the late 1980s and 1990s as well. Straightforward visions of the future – involving colonisation of the planets for the technophiles and optimists, or apocalypse of one kind or another for the technophobes and pessimists – are no longer so obvious or easy. Some recent science fiction novels – such as John Kessel's *Good News from Outer Space* (1989), Sheri S. Tepper's *Gibbon's Decline and Fall* (1996) or Elizabeth Hand's *Glimmering* (1997) – have taken refuge in an engagement with the Christian rhetoric of the fading millennium. Only Kim Stanley Robinson, in *Pacific Edge* (1990) and *Blue Mars* (1996), seems able to envision a utopian future; otherwise we are treated to endless decaying (but computerised) twenty-first-century cities, or far-flung futures with no connection to, or solutions for, the present. Christian prophecy and science fiction used to offer forms of certainty about the future: they represented, in a sense, crystallisations of the two extremities of popular understanding about what was in store for humanity. The changes we have seen in their visions in the last two decades of the century may thus represent a crystallisation of the inchoate changes in popular understanding, which some wish to sum up in the Sibylline ambiguity of the word 'postmodern'.

Notes

1. Doubt was expressed as early as the third century concerning the traditional identification of the author with the evangelist; most scholars now assume that these doubts were correct, and some acknowledge this with the sobriquet 'of Patmos', after the island on which the vision was said to have taken place.
2. As in the famous story by Arthur C. Clarke, 'The Nine Billion Names of God', in which, after the computer has listed all the names of God, the protagonists look up: 'overhead, without any fuss, the stars were going out'. First published in Frederik Pohl, ed., *Star Science Fiction Stories* (New York: Ballantine, 1953), and much anthologised since; reference from Clarke, *The Other Side of the Sky* (London: Gollancz, 1962), p. 12. See E. James, *Science Fiction in the Twentieth Century* (Oxford: Oxford UP, 1994), pp. 106–7.
3. Joseph Fielding Smith, *Doctrines of Salvation*, I (Salt Lake City: Bookcraft, 1954), pp. 88–89; quoted in A. A. Hoekema, *The Four Great Cults* (Exeter: Paternoster, 1963), p. 71.
4. W. S. Towner, in B. M. Metzger and M. D. Coogan, eds., *The Oxford Companion to the Bible* (New York: Oxford UP, 1993), p. 151.
5. On the role of the Antichrist in eschatological thought, see Bernard McGinn, *Antichrist: Two Thousand Years of the Human Fascination with Evil* (San Francisco: HarperSanFrancisco, 1994).
6. Quoted in B. McGinn, 'Early Apocalypticism: the ongoing debate', in C. A. Patrides and J. Wittreich, eds., *The Apocalypse in English Renaissance Thought and Literature* (Manchester: Manchester UP, 1984), 2–39, at 28.
7. D. Fergusson, 'Eschatology', in Colin E. Gunton, ed. *The Cambridge Companion to Christian Doctrine* (Cambridge: Cambridge UP, 1997), pp. 226–244, at p. 241.
8. Ibid. p. 232.
9. Norman Cohn, *The Pursuit of the Millennium: Revolutionary Millenarians and the Mystical Anarchists of the Middle Ages* (rev. ed., London: Temple Smith, 1970).
10. See, for instance, C. Hill, *Antichrist in Seventeenth-Century England* (rev. ed., London: Verso, 1990), and *The English Bible and the Seventeenth-Century Revolution* (London: Allen Lane, 1993).
11. Hill, *Antichrist* (1990), p. 177.
12. Hill, *Antichrist in Seventeenth-Century England* (1st ed., Oxford: Oxford UP, 1971), p. 159.
13. M. Reeves, *The Influence of Prophecy in the Later Middle Ages: A Study in Joachism* (Oxford: Oxford UP, 1969), p. 508.
14. Paul Boyer, *When Time Shall Be No More: Prophecy Belief in Modern American Culture* (Cambridge, Mass.: Harvard UP, 1992), p. 15.

15. Boyer, *op. cit.*, pp. 13–15, and notes on p. 345.
16. W. Warren Wagar, *Terminal Visions: The Literature of Last Things* (Bloomington: Indiana UP, 1982), p. 33.
17. This is discussed by Wagar, *op. cit.*, p. 21.
18. E. F. Bleiler, *Science Fiction: The Early Years* (Kent, Ohio: Kent State UP, 1990), p. 55.
19. Benson followed this with *The Dawn of All* (1911), which was intended as a Catholic utopia, but reads like 'a violent and damaging satire on Catholicism': Bleiler, *op. cit.*, p. 57. It is not in any sense a sequel to *Lord of the World*, as Clute claims in J. Clute and J. Grant, eds., *The Encyclopedia of Fantasy* (London: Orbit, 1997), p. 107.
20. *A Clash of Cymbals* (London, Faber and Faber, 1959), had been first published as *The Triumph of Time* (New York: Avon, 1958).
21. In the Faber edition, the date is on p. 175, the final words on p. 197.
22. It is unfortunate that Blish was persuaded to change the date of 4004 in later editions, because it conflicted with the chronology of the *Cities in Flight* sequence (of which this is the fourth volume). The problem is discussed by David Ketterer in *Imprisoned in a Tesseract: The Life and Work of James Blish* (Kent, Ohio: Kent State UP, 1987), pp. 184 and 347–48. Ketterer optimistically suggests that the revised date of 4104 'still expressed something of the same idea [of enclosure and containment] but less absolutely': p. 184.
23. This cliché has been described by Brian W. Aldiss as a 'Shaggy God' story: see Brian Stableford on 'Adam and Eve' in J. Clute and P. Nicholls, ed., *The Encyclopedia of Science Fiction* (London: Orbit, 1993), pp. 4–5.
24. Poul Anderson, *Tau Zero* (1970), cited from Coronet edition (London, 1973), p. 181.
25. Such fictions have been much discussed: see, e.g., Wagar, *op.cit.*; Brian Stableford in 'Man-Made Catastrophes in SF', *Foundation* 21 (June 1981), 56-85; Martha Bartter, *The Way to Ground Zero: The Atomic Bomb in American Science Fiction* (Westport, Conn., 1988); Paul Brians, in *Nuclear Holocausts: Atomic War in Fiction, 1895-1984* (Kent, Ohio: Kent State UP, 1987) and 'Nuclear war fiction for young readers: a commentary and annotated bibliography', in P. J. Davies, ed., *Science Fiction, Social Conflict and War* (Manchester: Manchester UP, 1990), 132–50; David Dowling, *Fictions of Nuclear Disaster* (London: Macmillan, 1987); and see under 'Disaster', 'End of the World' and 'Holocaust and After' in Clute and Nicholls, *op. cit*. Paul Boyer, *By the Bomb's Early Light: American Thought and Culture at the Dawn of the Atomic Age* (New York: Pantheon, 1985), is a wider study, of great relevance to this question.
26. *Gay Hunter* was originally published as by Lewis Grassic Gibbon, in 1934; a new edition was produced in 1989 by Polygon (Edinburgh).
27. Boyer, *op. cit.*, p. 321.
28. John Clute under 'Survivalist Fiction', in Clute and Nicholls, *op. cit.*, p. 1188.
29. 'Vintage Season' was published in *Astounding* (September 1946) as by Laurence O'Donnell: some reckon this to be a pseudonym of Moore

and her husband Henry Kuttner (and is thus reprinted as such in David G. Hartwell, ed., *The World Treasury of Science Fiction*, Boston: Little, Brown, 1989, 980-1018), while others think it was by Moore alone. Silverberg's reworking was published in *Isaac Asimov's Science Fiction Magazine* (March 1989), 108–74.

30. First published in Terry Carr, ed., *Universe 2* (New York: Ace, 1972), 41–51, and reprinted in, e.g., Silverberg, *Unfamiliar Territory* (London: Coronet, 1973).
31. First published in Damon Knight, ed., *Orbit 5* (New York: Putnam, 1969), and reprinted in N. Spinrad, *No Direction Home* (New York: Ace, 1975) and Spinrad, *The Star-Spangled Future* (New York: Ace, 1979), 227–59.
32. For what follows, I am heavily indebted to Paul Boyer, *Prophecy Belief*, esp. pp. 75ff.
33. Referred to by Boyer, *op. cit.*, p. 106: Oilar's book was published by Meador, of Boston.
34. Published in Pamela Sargent and Ian Watson, eds., *Afterlives: An Anthology of Stories about Life after Death* (New York: Vintage, 1986), p. 138.
35. See the novels *The Days of Noah* by M. R. DeHaan (Grand Rapids: Zondervan, 1963); *666* by Salem Kirban (Wheaton, Ill., Tyndale House, 1970), and *Beast: A Novel of the Future World Dictator* by Dan Betzer (Lafayette, La.: Prescott, 1985), discussed and quoted by Boyer, *op. cit.*, pp. 258–9.
36. Hal Lindsey, *The Late Great Planet Earth* (1970, cited henceforth from the first UK edition: London: Marshall Pickering, 1971), p. 176. In the first sentence he probably meant to say that they rejected the literality of much of Scripture.
37. The figures are derived from Boyer, *op. cit.*, p. 5.
38. Lindsey, p. 17.
39. Ibid., p. 25.
40. Ibid., p. 31.
41. Ibid., p. 137.
42. Ibid., p. 154.
43. Preachers quoted by Boyer, *op. cit.*, p. 242.
44. Fergusson, *op. cit.*, p. 227.
45. Stephen D. O'Leary, *Arguing the Apocalyptic: A Theory of Millennial Rhetoric* (New York and Oxford: Oxford UP, 1994): ch. 6, 'Hal Lindsey and the Apocalypse of the Twentieth Century', pp. 134–71.
46. O'Leary, *op. cit.*, p. 169.
47. As discussed by O'Leary, *op. cit.*, pp. 180–83.
48. New York, Bantam, 1981.
49. Quoted and discussed by O'Leary, *op. cit.*, p. 175.
50. Quoted ibid. p. 177.
51. A speech delivered on 17 September 1986, quoted in O'Leary, ibid. p. 186.

5

Edwardian Awakenings: H. G. Wells's Apocalyptic Romances (1898–1915)

PATRICK PARRINDER

> And I saw a new heaven and a new earth: for the first heaven and the first earth were passed away.
>
> (Revelation 21: 1)

The early, *fin-de-siècle* Wells needs no introduction as an apocalyptic writer. In *The Time Machine* (1895) the Time Traveller's journey into the future reaches its terminus on a cold and dying Earth darkened by a solar eclipse. The extinction of humankind and the eventual waning of the sun's heat are scientific prophecies made vividly real by the device of time travelling. Wells returned to the theme of a cosmic apocalypse in his short story 'The Star' (1897) and in *The War of the Worlds* (1898). Exemplifying Oscar Wilde's pairing of *fin de siècle* and *fin du globe*, the Martian invasion in the latter book is specifically associated with the turn of the century. The exact year is not given, but the last sentence of the opening paragraph suggests a probable date of 1901: 'And early in the twentieth century came the great disillusionment.'[1] The year of the Eloi and Morlocks in *The Time Machine* AD 802, 701, is also the first year of a new century.[2] The title of W. T. Stead's 1898 review-article on Wells's writings, 'The Latest Apocalypse of the End of the World', suggests how readily Wells's early work was associated with millennial anticipations.[3]

Anticipations (1901) and *The Discovery of the Future* (1902) were, indeed, the two non-fictional works with which he marked the turn of the century. These books have generally been understood as marking a turn in his own writings, from pessimism to optimism, from science to politics, and from romance to futurology or short-term forecasting. The change is a complex one, coinciding with

Wells's growing celebrity, with his membership of the Fabian Society, and with his emergence as a novelist of contemporary life. At the same time he continued to write both short stories and a series of lively, controversial and topical romances set in the future. These future romances, *The Food of the Gods* (1904), *In the Days of the Comet* (1906), *The War in the Air* (1908), and *The World Set Free* (1914), are now among his most unjustly neglected writings. Though inferior to Wells's most famous scientific romances, they are all in varying degrees apocalyptic in character, and they show the persistence and development of a characteristically Wellsian vision of apocalypse long after the turn of the century.

While the future revealed in *The Time Machine* is based on Darwinism and the Laws of Thermodynamics, the later Wells foresees an impending Last War, political collapse, and, beyond that, an era of unlimited expansion. With hindsight it becomes evident that the latter themes are nascent in *The War of the Worlds*, where the cooling of the solar system forces the Martians to 'carry warfare sunwards' (p. 4) in their quest for survival. As the narrator reflects in the Epilogue, their attack on the Earth has 'done much to promote the conception of the commonweal of mankind' (p. 300), and he foresees a second, decisive war between the Martians and a united humanity. The winners will be free to explore beyond the solar system, though the narrator admits that 'To them, and not to us, perhaps, is the future ordained' (p. 301). The 'future' open to a successful, intelligent species is thus one of world unification, space travel and galactic conquest. The Earth has been awakened to this future by a catastrophic invasion which 'in the larger design of the universe [...] is not without its ultimate benefit for men' (p. 300). In *The Time Machine*, the 'larger design of the universe' was, simply, one of extinction. The *War of the Worlds* introduces three other possible meanings of the phrase. The first is that of Christian fundamentalism. The second or Promethean interpretation suggests that by seizing its opportunities and by harnessing the forces of nature humanity can become the controllers of the universe. Thirdly, however, the future may be ordained 'to them, and not to us'. That is, either humanity will fail the test awaiting it, or it will win through only at the cost of transforming itself into a race of super-beings with which ordinary contemporary humanity will be unable to identify.

In nearly all of Wells's apocalyptic romances there are characters who understand catastrophic events as tokens of divine vengeance.

The curate in *The War of the Worlds*, for examples, quotes from the Book of Revelation and believes the Martians to be the harbingers of the Second Coming. Though the narrator tries to see the Martians in evolutionary terms and to play down their monstrous aspects, he too thinks of them as usurping the divine power: hence his satisfaction when they are brought low by terrestrial bacteria, 'the humblest things that God, in his wisdom, has put upon this earth' (p. 282). Later he is left 'weeping and praising God' (p. 288) in the aftermath of the Martian collapse. God himself appears as a character in 'A Vision of Judgment' (1899), a light-hearted fantasy (recalling Byron's similarly titled poem) that Wells wrote soon after *The War of the Worlds*. Here God and the Recording Angel are shown punishing the sinful by exposing them to the ridicule of their fellow human beings. Each soul is held in the palm of God's hand, but they are eventually allowed to scuttle away and hide up the divine sleeve. At the end, instead of consigning everyone to Hell God shakes them out of his sleeve onto a new planet in orbit around Sirius. The result of the Second Coming is that humanity gets a second chance, with instructions to 'try again' on a 'beautiful land, more beautiful than any I had ever seen before – waste, austere, and wonderful; and all about me were the enlightened souls of men in new clean bodies'.[4] This passage, down to the suspension points, is typical of the later Wellsian apocalypse in its optimistic and deliberately incomplete gesture towards a new life on a transformed earth.

Thanks to God's intervention, in 'A Vision of Judgment' the prospect of planetary cooling represented as a terminal horizon in *The Time Machine* has evidently lost its urgency. Wells wrote in *The Discovery of the Future* that solar entropy was 'of all nightmares, the most convincing', but then he immediately dismissed such fears, affirming his belief in the 'coherency and purpose in the world and in the greatness of human destiny'.[5] This apparently irrational profession of faith in a larger design was soon to receive a kind of scientific justification. Wells observed in a footnote to the 1913 reissue of *The Discovery of the Future* that the nightmare of solar entropy was no longer convincing since 'the discovery of radio-activity has changed all that'.[6] His chief source of knowledge about radioactivity and nuclear energy was Frederick Soddy's popular outline of *The Interpretation of Radium* (1908). According to Soddy, the new theory of the atom meant that:

> our outlook on the physical universe has been permanently altered. We are no longer the inhabitants of a universe slowly dying from the physical exhaustion of its energy, but of a universe which has in the internal energy of its material components the means to rejuvenate itself perennially over immense periods of time, [...][7]

If civilisation in Soddy's view could be summed up as a 'fight with Nature for energy' (p. 253), then 'possibilities of an entirely new civilisation are dawning with respect to which we find ourselves still on the lowest plane' (p. 239). Soddy's analogy between the dawn of the new civilisation and our ancestors' harnessing of fire was echoed by Wells in *The World Set Free*, where the opening section outlining the discovery of atomic energy has the Promethean title of 'The Sun Snarers'.[8]

Where *The Time Machine* has portrayed the 'Sunset of Mankind', many of Wells's later romances celebrate what is proclaimed as the dawn of history.[9] For example, *The Food of the Gods and How It Came to Earth* (to give the book its full, Promethean-sounding title) ends with young Cossar's vision of future beings conquering the heavens and growing 'into the fellowship and understanding of God till the earth is no more than a footstool'.[10] The novel is ambivalent about whether these future beings, the descendants of the Giants, would still be recognisably human. *In the Days of the Comet* has an epigraph from Shelley, 'The World's Great Age begins anew', and tells of a passing comet which infuses the atmosphere with green gases, bringing about human reunification and the construction of a better society.[11] Another vision of a 'new heaven and a new earth' from the same period in Wells's writing is *A Modern Utopia* (1905), set on a distant planet which is a physical replica of the Earth with, at this moment in history, a genetically identical human population. When the Utopian illusion fades and the narrator finds himself back in modern London, he briefly envisages an apocalyptic angel, 'a towering figure of flame and colour, standing between earth and sky, with a trumpet in his hands', calling for the world to awaken.[12]

The Promethean aspect of Wells's vision requires humanity to build a new world by its own efforts. Unfortunately, in the nineteenth century technology had far outstripped the political development of the major industrial nations. The result, as Wells's narrator summed it up in *The War in the Air*, would necessarily be the 'logical outcome of the situation created by the application of science to warfare' – in other words, global destruction.[13] After the inevitable world war, Wells foresees a profound change in human life, to be brought about either by some kind of external intervention or by the devastating impact of the war itself. Once the Last War involving aerial conflict has taken place, the present stage of history will be over, and in time the 'dream of a world's awakening' can be realised.[14] The vision of a Last War would inspire Wells, shortly after the outbreak of hostilities in August 1914, to produce his bestselling topical pamphlet *The War That Will End War*. The concept's apocalyptic origins are apparent in all his early twentieth-century romances, beginning with the short story 'A Dream of Armageddon'.

'A DREAM OF ARMAGEDDON' (1901)

'A Dream of Armageddon' is constructed on a triangular pattern with three forces in contention with one another. These are the belligerence of nation-states driving to war and destruction, the possibility of creative statesmanship overcoming national division, and the instinct for private fulfilment and self-preservation at the expense of public duty. The story takes the form of a prophetic dream of events at some indeterminate point in the future (Wells in a cancelled draft gave the date as 2100).[15] Apart from the advent of mass tourism and universal air transport, society has progressed very little during the 'long peace' that has presumably lasted since the beginning of the twentieth century.[16] Hedon, a leading statesman, has abandoned the political sphere in order to live openly with his mistress in the 'Pleasure City' of Capri. In his absence, the global situation rapidly deteriorates under his warmongering successor, Evesham. Hedon refuses every appeal to him to return to political life, and finally he and his companion are butchered, like thousands of others, in a conflict that he might have helped humanity to avoid. The story focuses on an ostensible conflict between eroticism and public service, but at its centre is Wells's deeply ambivalent use of the name of the Last Battle. Is Armageddon preventable by prompt and unselfish action, or is it part of the 'larger design of the universe' regardless of human decisions? Wells in 1927 acknowledged the power of the symbolism in this tale by putting it at the end of his *Complete Short Stories*.

THE FOOD OF THE GODS (1904)

The Food of the Gods begins as a comic version of the Frankenstein theme. Two bungling, unworldly researchers, Bensington and Redwood, rent a farm in north-west Kent in order to experiment with Herakleophorbia, a substance which overcomes the normal factors inhibiting growth in biological organisms. But they fail to take proper precautions against contamination, and breed not only giant children but hypertrophied plants and animals of all kinds. Before long, their friend Cossar has to lead a quasi-military expedition to burn the experimental farm and exterminate its population of giant wasps and rats. The farm's caretakers, Mr and Mrs Skinner, take to their heels, but Mrs Skinner carries off a tin of Herakleorphobia to feed to her baby grandson, in the village of Cheasing Eyebright. A generation later, young Caddles is shot down by a posse of riflemen, while the Cossar and Redwood children gather together in a fortified encampment built around the Cossars' family home in the North Downs. They are last seen preparing to defend themselves with Herakleophorbia-filled shells against the rage of the 'little people', led by the charismatic politician Caterham, who is determined to wipe them out.

Caterham's base is in London, while the giant children emanate from Wells's childhood county of Kent, which as he knew, was the home of John Ball and thus (arguably) the birthplace of socialism.[17] The war between ordinary humanity and the League of Giants is a kind of Peasants' Revolt in reverse. *The Food of the Gods* was clearly intended as a political allegory – Wells later described it as a 'fantasia on the change of scale in human affairs' – and the giants, representing the future, are ranged against the reactionary populist armies of modern democracy.[18] The book ends with the vision of young Cossar caught in the searchlights surrounding the giants' encampment at night. He is a 'great black outline that threatened with one mighty gesture the firmament of heaven and all its multitude of stars' (p. 287). But if the giant children can foresee a future of galactic conquest, their alienation from earlier humanity, is expressed not only by Caterham and his followers but by Redwood, the inventor who recoils from his creatures even though some of them are his own children. 'We have made a new world, and it isn't ours. It isn't even – sympathetic' (p. 277), Redwood complains. In fact the forty-foot-high giants are reminiscent not only of Gulliver in Lilliput but of the Martians mounted on their gigantic Fighting Machines.

IN THE DAYS OF THE COMET (1906)

Wells's next romance, written to anticipate the return of Halley's Comet in 1909, describes the world before and after the miraculous 'Change' brought about by the comet's near miss. Here he portrays a sense of estrangement between present and future humanity, rather than between humans and apparently superhuman beings. Continuity between the old world and the new is represented in the person of the narrator, Willie Leadford, who looks back on his youth in the feverish, hysterical last days before the Change. War was breaking out between England and Germany, there was bitter industrial conflict, and Ladford, maddened by class hatred and sexual jealousy, was on the point of murdering his rival in love. He is initially seen, as in a dream, sitting in a tower with high windows overlooking a utopian city full of 'galleries and open spaces, ... trees of golden fruit and crystal waters' (p. 305). This vision of the 'Happy Future' recalls the outer narrator of Henry James's earthly paradise, the Great Good Place (p. 3). The outer narrator can easily understand Leadford's account of his early life, though Leadford fears that his experiences will be 'inconceivable', as if written in an 'unknown language', for those born after the Change (p. 114). It is the world after the Change that the outer narrator finds it barely possible to come to terms with:

> I had been lost in his story throughout the earlier portions of it, forgetful of the writer and his gracious room, and the high tower in which he was sitting. But gradually, as I drew near the end, the sense of strangeness returned to me. It was more and more evident to me that this was a different humanity from any I had know, unreal, having different customs, different beliefs, different interpretations, different emotions. It was no mere change in conditions and institutions the comet had wrought. It had made a change of heart and mind. In a manner it had dehumanized the world, robbed it of its spites, its little intense jealousies, its inconsistencies, its humour. At the end [...] I felt his story had slipped away from my sympathies altogether. [...]. As the change was realized, with every stage of realization the gulf widened and it was harder to follow his words. (pp. 303–4)

Here the difficulty of narrating one world to the other denotes their mutual estrangement: the languages are different, and the words cannot be followed. Wells had written elsewhere of the inevitable effect of 'hardness and thinness' in utopian speculations;[19] here – perhaps anticipating the scandalised reaction to his novel's sexual morality – he implies that a utopian world must seem, to some degree, 'dehumanized' to its readers.

In the Days of the Comet differs from Wells's other romances in that the Change is the result of a cosmic accident, owing nothing to human technological achievements. The Promethean element is thus absent from the book. In the opening chapters, the industrial setting in the Potteries suggest a degree of fictional realism which is belied by the hero's emotional turmoil and increasingly melodramatic behaviour. Not surprisingly, religious ideas and apocalyptic images are prominent in the text. Leadford hears a street-corner preacher linking the comet's approach to the prophecies of the Book of Daniel (p. 72). He has earlier described the night visions caused by the appearance of the comet and the polluted 'dust-laden atmosphere' (p. 28), and he later speaks of the 'great monstrous shapes of that extraordinary time' (p. 97); these may be taken as references to the coming of the kingdom of the saints in Daniel Chapter 7. At the culmination of this 'mad era' (p. 98), when he is about to open fire on the eloping lovers at their hideout on the east coast, Leadford claims that 'The war below, the heavens above, were the thunderous garment of my deed' (p. 171). But the Last Days are suddenly brought to an end by the comet's vapours, which temporarily paralyse all living things. Leadford's first impression when he regains consciousness is that he is in the 'barley fields of God' (p. 179):

> The imaginations of my boyhood came back as speculative possibilities. In those days I had believed firmly in the necessary advent of a last day, a great coming out of the sky, trumpetings and fear, the Resurrection, and the Judgement. My roving fancy now suggested to me that this Judgement must have come and passed (p. 183).

What has come about, we soon discover, is the dawn of rationality. Soon (in some of Wells's least likely scenes) Leadford watches the leading politicians of the time coming together and confessing their errors and follies before convening a conference to end the war and set up the World State. Among the common people, the Change finds expression in a religious revival and, for many, the comet's approach is taken for the 'Second Advent' (p. 221). Leadford's attitude to such beliefs is studiously agnostic, though on at least one occasion he prays to his own image of a 'Master Artificer, the unseen captain of all who got about the building of the world' (p. 268). The main communal celebration of the Change is an annual quasi-religious ritual, the 'Beltane fires' which recognise that destruction must come before reconstruction. In these 'Phoenix fires' (p. 288) the cultural debris of the old society, including its books, is systematically burnt. Cities too are swept away, with London and Oxford being replaced by 'Caerlyon and Armedon, the twin cities of lower England' (p. 238).

The young Leadford is a rude and arrogant convert to socialism and the gospel of equality. When, immediately after the Change, he comes upon the statesman Melmount, who has sprained his ankle, Leadford instantly becomes humble and deferential. His 'inflamed, [...] rancid egotism' has gone (p. 192). Here Wells implies that false aristocracy of our present class society would give way under utopian conditions to a rational aristocracy of leaders and led. (The name Leadford can itself be understood in this light.) Once Willie's true capacities have been released by the Change we see him as a natural civil servant, happy to be sent back to the Potteries to 'help prepare a report' for the new government (p. 246), and then to join the provincial administration.

THE WAR IN THE AIR (1908)

The next romance portrays the breakdown of civilisation through the adventures of an even more humdrum hero, Bert Smallways, who resembles the Wellsian comic heroes Kipps and Mr Polly. Bert is first shown helping to run a bicycle repair shop in Bun Hill, a small town south of London evidently modelled on the Bromley of Wells's childhood. It is a time of ballooning, monorails, and the first aeroplanes, but all that can be afforded by Bert, the 'progressive' member of the Smallways family (p. 6), is a dodgy motor-cycle. When this machine blows up in the course of a country excursion, Bert is seen as a 'sad, blackened Promethean figure, cursed by the gift of fire' (p. 51). He is in reality no Prometheus, but at best a witness and victim of technological progress.

Like its predecessor, *The War in the Air* is set at the 'opening of the twentieth century' (p. 337), when science has rendered the 'old separations into nations and kingdoms' obsolete and a 'new, wider synthesis' is 'imperatively demanded' (p. 97). But within the narrative frame there are no signs of any new synthesis, nor is there a comet on hand to save the world. New York and its other great cities are devastated by aerial bombing, and Britain and Europe are shown returning to the Dark Ages. The narrator's Gibbonian standpoint, looking back on the early twentieth-century 'great collapse' (p. 336) as an analogue of the fall of Rome, implies that the eventual outcome will be the rise of a new civilisation.[20] A single allusion to 'our present world state, orderly, scientific, and secured' (p. 336) confirms that this is the case. The *War in the Air* thus resembles *The Shape of Things to Come* (1933) and its film version *Things to Come* (1936), where the Last War leads to decades of anarchy as a necessary prelude to the construction of the world state. The prominence of Bert Smallways's adventures and the ending of the story while 'Anarchy, Famine, and Pestilence' are still triumphant (p. 346) has tended to obscure the links between *The War in the Air* and these later exercises in future history.

Once again, the War in the Air is an apocalyptic event for those who witness it at first hand. The fleet of airships attacking New York, the 'modern Babylon' (p. 172), is 'like a flock of strange new births in a Chaos that had neither earth nor water, but only mist and sky' (p. 171). The German war leader, Prince Karl Albert, tells his followers that they are bringing a Teutonic twilight of the gods: 'Fifty centuries come to their Consummation', he boasts (p. 225). A group of 'Christianized half-breeds' in Labrador debate whether or not his crippled flagship is a sign of the Second Coming (p. 219), while to Kurt, the German lieutenant, the news of global warfare represents 'the end of the world' (p. 229). Bert Smallways too believes that he is witnessing 'the sunset of his race' (p. 270). With hindsight, these expectations fall into historical perspective: humanity has bombed itself not to oblivion but back into the agricultural stage of society, represented by Bert's brother Tom Smallways, who begins the novel as a gardener and greengrocer and ends up as a small farmer. Bun Hill by mid-century has 'rediscovered religion and the need of something to hold its communities together', and this is supplied by an old Baptist minister who warns his flock against Alcohol and the Scarlet Woman (p. 361). In time, we must assume, the minister's rantings will be replaced by the revelations of a more rational, 'orderly' and 'scientific' faith.

THE WORLD SET FREE (1914) AND 'THE STORY OF THE LAST TRUMP' (1915)

The World Set Free, now mainly remembered for its anticipation of the atomic bomb, is much less apocalyptic in tone than its predecessors. Instead, an orderly, scientific faith prevails, and the romance is the poorer for it. Nor does it have a colourful figure like Leadford or Bert Smallways at its centre. The narrator, a future historian, summarizes at some length the memoirs of a young officer, Frederick Barnet, but Barnet is a narrative convenience of no intrinsic interest. Though he at one point describes the atomic bombs (thrown out of aeroplanes) as falling 'like Lucifer in the picture' (p. 123), he is a 'Modern State' man 'by instinct' (p. 55), untroubled by millennial anticipations.

The discovery of nuclear energy is a Promethean event, opening up the possibility of an entirely new civilisation transforming the Earth and leading to the conquest of space (pp. 27–8). In Wells's novel commercial nuclear power generation begins in 1953, causing a second industrial revolution which brings about mass unemployment. This quickly leads to the outbreak of war. The early stages of the war are rather presciently imagined (the novel was published in may 1914), but the atomic explosions soon bring the world's leaders to their senses. A conference is called, the unregenerate King of the Balkans is ambushed and executed, and the World State is established. W. Warren Wagar, for one, has contrasted this simplistic denouement with the far more daring (and more pessimistic) presentation of humanity's decline in *The War in the Air*.[21]

The insistence that human development up to the present time is only a 'beginning' (pp. 20, 266) resounds throughout the expository passages of *The World Set Free*, just as it would recur in the Finale of *Things to Come* twenty years later.[22] The new phase in history evoked in *The World Set Free* is post-Christian, recognising Christianity as the 'first expression of world religion' but leaving behind its 'temporary forms'; revelation and the Second Coming have given way to a Promethean faith in science and the 'common sense of mankind' (p. 243). But this is by no means the last of the Wellsian apocalypse. *The World Set Free* refers to the present as the 'womb of the future', (p. 105), and it ends with the death of an old seer, in a chapter called 'The Last Days of Marcus Karenin' (pp. 244–86).

Wells's later intimations of an impending end to all things could be traced down to his despairing last book, *Mind at the End of Its*

Tether (1945), written shortly before his own death. This essay will conclude with an apocalyptic allegory published early in the First World War which shows him in full reaction against the optimism of *The World Set Free*. In 'The Story of the Last Trump' (1915), the trumpet that God has stored up to blow at the Last Judgement is found in a celestial attic by a mischievous child, who inadvertently knocks it over the battlements. It falls to Earth and languishes in an antique shop until two apprentices come along and decide to play every musical instrument in the shop for a bet. The trumpet resists all attempts to sound it until one of the young men ingeniously connects it to a factory blow-pipe. The instrument breaks into pieces, none of which are ever found, but all over the world its sound is heard and then instantly cut short. Needless to say, humanity fails to obey the summons. 'Men will go on in their ways as rabbits will go on feeding in their hutches within a hundred yards of a battery of artillery,' Wells concludes.[23] He is himself both the mischievous child and the resourceful apprentice – the vehicle of an apocalyptic prophecy, uttering a premature summons to a world that obstinately refuses to awaken. Time and again, Wells would draw upon the religious imagery of the end of the world in his search for a language powerful enough, and urgent enough, to convey his sense of human frailty and human destiny.

Notes

1. H. G. Wells, *The War of the Worlds* (London: Heinemann 1898), p. 2.
2. 802, 701 is 1901 plus 800,800 years. For the significance of this, see Patrick Parrinder, *Shadows of the Future* (Liverpool: Liverpool University Press, 1995), pp. 42, 73–5.
3. W. T. Stead, 'The Latest Apocalypse of the End of the World', *Review of Reviews*, 17 (1898), pp. 389–96.
4. H. G. Wells, *The Complete Short Stories* (London: Benn, 1927), p. 114.
5. H. G. Wells, *The Discovery of the Future with The Common-Sense of World Peace and The Human Adventure* (London: PNL Press, 1989), p. 34.
6. Ibid., p. 37.
7. F. Soddy, *The Interpretation of Radium* (3rd edn. London: Murray, 1912), p. 248.
8. H. G. Wells, *The World Set Free: a Story of Mankind* (London: Macmillan, 1914), pp. 1–29. The book is dedicated to Soddy's *Interpretation of Radium*.
9. H. G. Wells, *The Time Machine: An Invention* (London: Heinemann, 1895), title of Chapter 6.

10. H. G. Wells, *The Food of the Gods and How It Came to Earth* (London, Nelson, n.d.), p. 286.
11. H. G. Wells, *In the Days of the Comet* (London: Macmillan, 1906).
12. H. G. Wells, *A Modern Utopia* (London: Chapman & Hall, 1905), p. 369.
13. H. G. Wells, *The War in the Air* (London: Nelson, n.d.), p. 202.
14. H. G. Wells, *A Modern Utopia*, p. 369.
15. J. R. Hammond, *H. G. Wells and the Short Story* (Basingstoke: Macmillan, 1992), p. 144.
16. H. G. Wells, *The Complete Short Stories* p. 1026.
17. H. G. Wells *'42 to 44': A Contemporary Memoir* (London: Secker & Warburg, 1944), pp. 27–30.
18. H. G. Wells, 'Preface to *The Scientific Romances*', in P. Parrinder and R. M. Philmus (eds.), *H. G. Wells's Literary Criticism* (Brighton: Harvester, 1980), p. 243.
19. H. G. Wells, *A Modern Utopia*, p. 9.
20. On Wells and Gibbon see P. Parrinder, *Shadows of the Future*, pp. 65–79.
21. W. Warren Wagar, *Terminal Visions: The Literature of Last Things* (Bloomington: Indiana University Press, 1982), p. 122.
22. H. G. Wells, 'Things to Come' in *Two Film Stories* (London: Cresset, 1940), p. 141.
23. H. G. Wells, *The Complete Short Stories*, p. 604.

6

Acts of God

ROBERT CROSSLEY

It looks like the end of the world – or like any number of twentieth-century episodes in extermination. Gigantic, mobile machines of war vaporize country railway stations. Smoke rises above scorched ground where lately stood comfortable suburban houses. Refugees fill the roads and crowds of corpses lie unburied. A tranquil landscape has become a living hell. 'What do these things mean? ... Why are these things permitted? What sins have we done?'[1] The questions asked by H. G. Wells's agitated curate in *The War of the Worlds* are familiar human plaints in the face of unspeakable catastrophe. The curate's desperate desire for an adequate interpretation of the initial deaths and mayhem caused by the Martian invaders in the suburbs of London, his querying of divine acquiescence in and human responsibility for the disaster are the distinctive human responses to catastrophe from the early modern era to the present.

If the eminent historian of plagues, William H. McNeill, is correct that the great writers of the Middle Ages all treated the devastation of the Black Death in the fourteenth century as 'a routine crisis of human life – an act of God, like the weather',[2] then one may get a glimpse of the shift toward the distinctively modern questioning of God's acts in Shakespeare's most apocalyptic play. The final scene of *King Lear* is full of allusions to the cataclysmic events of the end of the world, but its most piercing accusation against divine indifference and hostility to human suffering follows immediately on the Duke of Albany's prayer that Cordelia's life be spared ('The gods defend her'). The plural and lower-case 'gods' casts only a threadbare and transparent veil over the play's indictment of divine providence. Shakespeare's accusation occurs in the form of a stage direction and in the animal cries torn out of Lear as he carries his child's corpse:

> Enter Lear, with Cordelia in his arms.

> LEAR. Howl, howl, howl, howl! O, you are men of stones:
> Had I your tongues and eyes, I'ld use them so
> That heaven's vault should crack.

In the four centuries since *King Lear* was first performed God's interventions and, just as importantly, God's failures to intervene, in human affairs and natural processes has called forth a variety of eloquent protests. Voltaire saw the devastation of Lisbon in the earthquake of 1755 as pointing to an inexplicably cruel or a non-existent God, and two centuries later Karen Gershon was still more scathing on the silencing of faith by the smell of 'the gas of Auschwitz on God's breath'.[3] Just twenty years after Wells's fictional clergyman questioned the meaning of the Martian invasion, an actual Methodist minister surveyed the ruins of Halifax, Nova Scotia, after one of the worst disasters ever recorded in North America. On 6 December 1917 two ships, one a French munitions freighter, collided in Halifax harbor and the resulting explosion leveled the city and caused 1,700 deaths – including those of the minister's wife and son. In his grief the Reverend William Swetman declared, 'If I thought this was an act of God, I would tear off this collar.'[4]

What most differentiates the outbursts of Voltaire, Gershon, and Swetman from those of the curate in *The War of the Worlds* is that Wells chose to depict his cleric as a figure of farce. In *The War of the Worlds* the more eschatological the curate's outbursts become, the more absurdly they are rendered:

> This must be the beginning of the end! ... The end! The great and terrible day of the Lord! When men shall call upon the mountains and the rocks to fall upon them and hide them – hide them from the face of Him that sitteth upon the throne!

The curate's hysteria prompts a reprimand from the narrator, who has, to his regret, befriended him. The scolding comes, surely, as much from Wells himself as from the narrator, whose contempt for the curate's parochial brand of faith is evident:

> What good is religion if it collapses under calamity? Think of what earthquakes and floods, wars and volcanoes, have done before to men! Did you think God had exempted Weybridge? He is not an insurance agent.[5]

The characteristic Wellsian satire on religious sentimentality and unreason is, despite its effort to amuse, not very different in implication from those questions about the nature and operation of the divine will that have been wrenched out of the cataclysms of history.

The inventory of catastrophic occurrences listed by Wells's narrator – earthquake, flood, volcano, war – is not only the stock in trade of apocalyptic literature, it also duplicates the items in standard insurance policy disclaimers of liability for damages: that collection of occurrences known technically as 'perils' or 'acts of God'. But the simple old phrase 'acts of God' labels with apparent definitiveness precisely what is most ambiguous in the interpretation of calamities on the grand scale: are they simply 'natural' occurrences? are they the products of human error or malfeasance? are they supernatural interventions into human and natural order? While insurance policies exempt hurricanes, earthquakes, tornadoes, floods and other meteorological events as products of natural causes without any human intervention, the status of these events as 'acts of God' in that narrow sense may be debatable. The coastal flooding that is likely to occur throughout the world in the twenty-first century because of climatic changes triggered by increased human generation of carbon dioxide is a salient example of the problem of interpreting an insurer's 'act of God'.

The phrase 'act of God' is a medieval renaming of the ancient Roman legal concept of *vis major* – a superior force operating as an extenuating factor that mitigates or excuses an obligation. But when English scholars began substituting the Christian and vernacular formula of 'act of God' for the old Latin phrase, legal authorities were quick to rule that, despite its apparent implication, 'an act of God did not depend on divine influence'. Differentiating between theological and mercantile applications, judges insisted that an act of God did not imply Almighty responsibility for catastrophes, but merely designated an unpredictable phenomenon of nature unassisted by human agency.[6] As a legal term, 'Act of God' is defined by the *Oxford English Dictionary* to mean an 'action of uncontrollable natural forces in causing an accident, as the burning of a ship by lightning'.

However, if the legal history of the term leaves little doubt that 'act of God' has had a clearly secularized meaning from the beginning, the uses of acts of God in apocalyptic literature have been far more various, rich in irony, pointedly ambiguous. Much of that rich tradition stems from the Biblical account of the first time the world ended: 'The end of all flesh is come before me,' says the God of Noah, 'for the earth is filled with violence through them; and, behold, I will destroy them with the earth' (Genesis 6: 13). Stories of the Great Deluge, retellings and reinterpretations of the events in chapters 6 through 8 of Genesis, furnish revealing examples of how acts of God have raised questions about both human and divine culpability for catastrophic events.

In the Chester play of Noyes Fludde, the text on which Benjamin Britten based his opera, Noah takes 120 years to build the ark; his dilatory carpentering is explicitly intended to buy time for the human race in the unfulfilled hope that divine mercy will eventually overcome God's grievance against his creation. When that strategy fails, Noah identifies himself with his Lord's act of destruction and becomes, in effect, his agent: 'Ah! great God that art so good. / He that does not thy will is but a clod.'[7] The most intriguing character in the Chester play is Mother Noah, who protests not merely the destructive act of God in ending the world but the selective exemption granted to her husband and his family. She will not voluntarily be separated from her sisterly drinking companions, her 'gossips', who plaintively describe the coming of the flood and their fear of drowning. The role of Noah's wife is larger than the conventional reading of her as comic shrew and tippler; she stands with humanity, with the human community, against the divine edict. Choosing to drown with her women friends, she enters the ark only when she is dragged there by her three sons on her husband's orders. She repays this act of salvation with a good punch.

From the possibilities sketched in the medieval mayhem in this dramatic conception of Noah and his wife, Timothy Findley has constructed a Mrs Noyes in his 1984 novel *Not Wanted on the Voyage* who, in deed and word, is a relentlessly acute critic of both a doddering Yahweh and her godly husband Dr. Noyes. This is a rebellious and fiercely loving Mrs Noyes, whose gin-drinking is both homage to her medieval prototype and symptom of the stresses of her lonely stand on behalf of the human victims of divine acts and divinely sanctioned human acts of malice and destruction. Her grandest and most chilling moment comes after her husband, in the name of God, has one of his sons kill a retarded child, Lotte, whom she has brought on to the ark but whom Noah deems only an ape, not a human. Lotte is 'not wanted on the voyage', not fit for salvation. When Mrs Noyes receives the body of

Lotte, she stands aboard the not-yet-launched ark, in the pouring rain, lifts the child to the sky, and makes her atheist's testament: '"There is no God," she said. "There is no God worthy of this child. And so I will give her back to the world where she belongs."'[8]

There is a more oblique reworking of the Noah story in John Calvin Batchelor's *The Birth of the People's Republic of Antarctica.* Much of the novel is a catalogue of late-twentieth-century disasters, political, geological, meteorological, medical, and moral as experienced by the protagonist, Grim Fiddle as he steadily moves south to the bleak landscapes of the South Atlantic and Antarctica. His grandfather, the Lutheran fundamentalist minister Mord Fiddle, escapes political calamity in Sweden aboard a modern ark called 'The Angel of Death'. Mord Fiddle annexes the Noah story to his own situation and uses it to catechize his agnostic grandson. Recalling how the God of Genesis found humanity 'loathsome' and intended the destruction of the earth, he puts to Grim Fiddle the critical question that lies at the heart of so many modern interrogations of apocalyptic theology, 'I ask, then, was Lord God a monster?'[9]

That question may be put to one of the most durable, and on its face one of the most theologically orthodox, science fiction apocalypses of the past fifty years.[10] Although for much of its length Walter Miller's *A Canticle for Leibowitz* seems centered on the question of human responsibility for catastrophe (the cyclical view of history, the repeated references to Eden and original sin, the recurrent conflicts between the moral imperatives of Catholicism and the political imperatives of secular institutions), near the end of the novel there is a powerful turn to questions of divine culpability. That turn is signaled by a surprising question posed by the central character in Part III, Abbot Zerchi, a figure of otherwise unbendingly doctrinaire conservative theology. As events in the thirty-eighth century move inexorably toward a second nuclear 'flame deluge' like the one that wrecked civilization in the twentieth century, Zerchi asks a younger monk: 'Is the species congenitally insane, Brother? If we're born mad, where's the hope of Heaven?'[11] Intimations that humanity is 'helpless', 'doomed', the prisoner of its God-given biology echo throughout the final chapters. Zerchi tries to maintain his pious belief in a just God and a wicked humanity, but two women characters offer formidable counterviews.

The first is an unnamed, dying young mother, with her radiation-blasted daughter, whom the abbot attempts to dissuade from suicide. Offered a chance at euthanasia for herself and her child, the mother is urged by Zerchi to suffer and accept as God's will the slow, painful deaths that lie ahead. When persuasion fails, the abbot resorts to bullying – he calls it 'adjuration' – and she nearly gives in to his commands until a policeman at the site of the crematorium requires Zerchi to release her and let her make her own free choice. Unrepentant, she chooses mercy-killing for herself and her daughter. The monk relies on detailed theological argument about the nature of pain and a protracted autobiographical account, at once graphically brutal and sentimental, of how as a child he killed a wounded cat to put it out of its misery. But the mother's disbelief carries more weight in its flat terseness: 'I cannot understand a God who is pleased by my baby's hurting.'[12]

Miller has an even more eloquent spokeswoman for this counterview in the mutant, two-headed old woman who peddles tomatoes at the monastery. On the day the world ends Mrs Grales goes to confession to Abbot Zerchi and startles him by asking forgiveness for her own sins and for God's sins. When she tells her confessor that she seeks 'shriv'ness' for 'Him who made me as I am', the monk is repulsed and reminds Mrs Grales that God is Love and God is Justice. But knowing God through his acts rather than as a compendium of theological abstractions, Mrs Grales nevertheless persists, 'Mayn't an old tumater woman forgive Him just a little for His Justice?'[13] Miller's narrative is one of the most striking instances of how the literary representation of apocalypse becomes a locus for theological inquiry and for spiritual self-scrutiny. Here the ante is raised on the old truism about the 'mysterious ways' in which God acts. In science fiction, in particular, imagined apocalypses are the high-profile occasions on which the conflicts between the scientific and the religious imaginations have been most frequently and intensely joined.

The question of whether God deserves responsibility – and perhaps might even be granted forgiveness – for his destructive acts is crucial to many apocalyptic narratives. In what W. Warren Wagar has named 'the first major example of secular eschatology in literature', *The Last Man*, Mary Shelley has an astronomer whose wife and children have died from plague curse God as 'the Supreme Evil' and renounce both the desire for heaven and the dread of eternal pain: 'I do not fear His hell, for I have it here.'[14] At a later stage of the global epidemic and in more moderate tones, the titular hero of the novel, Lionel Verney, asks a series of questions that evoke

Biblical promises overlaid by Hamlet's depressive rejection of comforting imagery: Did God create man, merely in the end to become dead earth in the midst of healthful vegetating nature? Was he of no more account to his Maker than a field of corn blighted in the ear? Were our proud dreams thus to fade? Our name was written 'a little lower than the angels'; and behold, we were no better than ephemera. We had called ourselves the 'paragon of animals,' and, lo! we were a 'quint-essence of dust'. With an eloquence more practiced, and perhaps finally less riveting, than that of Miller's old tumater woman, Verney struggles to adopt the accepting mood of Pope's deistic piety that 'whatever is is right'. Convinced that the dice are loaded against humanity, he nevertheless resolves that he 'will sit amidst the ruins and smile'.[15]

Shelley's questions predict those that arise in the most extensive fictional chronicle of disasters that has yet appeared, Olaf Stapledon's *Last and First Men,* which records two billion years' worth of things to come, including the rise and fall of eighteen different human species. In Stapledon's future history the end of the world, in fact, occurs a number of times, when human beings commit acts that make their own planet uninhabitable for many thousands of years, when they suffer invasion from Mars, and when cosmic forces destroy the human habitat on Earth and force migration first to Venus and ultimately to Neptune. Remarkably, some remnant of humanity survives all these disasters. While some of the disasters which end the various human civilizations are clearly the product of human vice or human tampering with natural processes, a number of them are, in the technical sense, 'acts of God'. Human history, the narrator informs us, 'has ever been "precarious" and at any number of points humanity just misses total extermination, until the final reckoning on Neptune two billion years from now'.[16]

Many of the events that initiate the endings of these multiple human cultures are characterized as 'accidents'. In fact, the very opening of *Last and First Men* suggests that by their nature human beings are accident-prone because of a fundamental inadequacy of the human nervous system to handle the stresses and demands of complex civilization. 'Animals that were fashioned for hunting and fighting in the wild were suddenly called upon to be citizens, and moreover citizens of a world-community.'[17] The passive voice leaves unspecified the fashioner of human physiology. An agnostic, Stapledon did not assume a God acting on the world, although in *Last and First Men* some of the imagined future cultures do develop mythologies and theologies that attribute agency to a deity. The Third Men, for instance, believe in an artist-god who persistently alters, reworks, and tortures his creation into new forms – with a streak of sadism that mirrors the Third Men's own exercises in perverse surgery on subordinate creatures.

The chapter titled 'Behaviour of the Condemned', near the end of the chronicle, is as close as *Last and First Men* comes to protesting an act of God. The Eighteenth (and last) Men, highly sophisticated descendants of the original human stock now living on Neptune, accept with serenity the prospect of the death of their own culture. They have developed a spiritual aesthetic that grasps the fluctuations of the cosmos as a symphony that can absorb the tragedy of any individual or culture and incorporate it into the music of the spheres. They had, we are told, 'even savoured in imagination the sudden destruction of our world'.[18] But when a stellar 'infection' of our solar system is detected, it becomes evident to the Eighteenth Men that the sun will soon go into the nova state and all life in the system will be sterilized. A project to propel the 'seeds' of human life into other uncontaminated systems is undertaken, although it is given small hope of success. This seems truly to be the last chapter in the human story. In the final days of their existence, the Eighteenth Men realize that the end of their culture is likely indeed to be the dead end of the human inheritance entirely, and their calm acquiescence collapses as a vision of a mad God emerges:

> We now see our private distresses and the public calamity as merely hideous. That after so long a struggle into maturity man should be roasted alive like a trapped mouse, for the entertainment of a lunatic! How can any beauty lie in that?[19]

The question of the place of pain in the cosmic design and the kind of deity that presides over the universe re-emerges in many of Stapledon's later fictions, but its most emphatic and definitive treatment is in *Star Maker*. In a mixture of plain truth and disingenuousness Stapledon described *Star Maker* as 'an imaginative sketch' of ultimate realities.[20] Less a chronicle than *Last and First Men* and more like an intellectual pilgrimage, *Star Maker* is organized as a linked series of philosophical vignettes in which the pilgrim-narrator observes so many apocalypses throughout the cosmos, throughout time, and throughout alternative times and cosmoses that speculation about the Creator's inexplicable purposes in destroying the worlds He has made becomes the central preoccupation of the narrative. The pilgrim's own philosophical dilemma mirrors the doubts and anxieties he finds expressed in all the 'minded' worlds he surveys. 'Blind chance, it was rumoured, ruled all things; or perhaps a diabolic intelligence. Some began to conceive that the Star Maker had created merely for the lust of destroying.'[21] Some of the civilizations visited by the pilgrim conclude, as do the tellers of the Biblical stories of the Deluge, of Sodom and Gomorrah, and of the Babylonian Captivity, that their ordeals are inflicted by divine wrath for impiety. In one of the boldest imaginings in *Star Maker* we are taken back to the universe's beginnings, into the minds of the primeval nebulae – the titanic first intelligences – and are invited to experience their apocalyptic despair as they undergo their own deaths in the birth-pangs of their transformation into stars. The nebulae sense that 'the cosmos was a place of futility and horror', and as each nebula expires it becomes evident to the dwindling survivors that the strange plague, as they conceived of the 'sickness' that was devastating them, 'was no casual accident but a fate inherent in the nebular nature'.[22]

The nebular vision is a paradigm for the dark-bright revelation that comes to every succeeding Stapledonian creature in the cosmic panorama sketched in *Star Maker*. Typically, disaster comes by accident or natural evolution within an order of creation – an act of God in the legal sense. There are civilizations and worlds that are annihilated in a planetary collision, or asphyxiated from the gradual loss of atmosphere, or frozen by the loss of heat from a dying sun. Other acts of God are in the insurer's problematic zone: apocalypses that result from intelligent species' tampering with natural processes or from interplanetary wars. But on occasion the pilgrim-narrator witnesses cultures eradicated through an act of God in the strictest theological sense:

> In general the Star Maker, once he had ordained the basic principles of a cosmos and created its initial state, was content to watch the issue; but sometimes he chose to interfere, either by infringing the natural laws that he himself had ordained, or by introducing new emergent formative principles, or by influencing the minds of the creatures by direct revelation.[23]

Stapledon introduces the climactic visionary sequence in *Star Maker* with a literal apocalypse: 'But now the veil trembled and grew half-transparent to the mental vision.'[24] The pilgrim at last perceives the Star Maker neither as loving parent nor as divine insurance agent but as a calculating artist, making decisions about creation and destruction, revision and obliteration solely on aesthetic grounds. An act of God, as Stapledon imagines it, is an act taken in ruthless pursuit of perfection, a pursuit untempered by affection for creatures, the pursuit rather of pure creativity. In the gaze that the Star Maker casts on his creations, the pilgrim observes a profoundly inhuman detachment:

> Here was no pity, no proffer of salvation, no kindly aid. Or here were all pity and all love, but mastered by a frosty ecstasy. Our broken lives, our loves, our follies, our betrayals, our forlorn and gallant defences, were one and all calmly anatomized, assessed, and placed. True, they were one and all lived through with complete understanding, with insight and full sympathy, even with passion. But sympathy was not ultimate in the temper of the eternal spirit; contemplation was. Love was not absolute; contemplation was. And though there was love, there was also hate comprised within the spirit's temper, for there was cruel delight in the contemplation of every horror, and glee in the downfall of the virtuous. All passions, it seemed, were comprised within the spirit's temper; but mastered, icily gripped within the cold, clear, crystal ecstasy of contemplation.[25]

Stapledon still found it possible to 'worship' (though not to love) such a divine spirit, but later apocalyptic writers have adopted his indictment of God's acts without endorsing his surrender of humanism to the adoration of cosmic mystery. Hence the defiant atheism of Findlay's Mrs Noyes and the fierce protest of Batchelor's Grim Fiddle against a 'God of Hate who had made mankind wretched the more to toy with and torture the sinner.'[26]

A writer who in many ways has been Stapledon's most popular disciple, Arthur C. Clarke, also carries forward this vision of a God without a human face or human empathy. But Clarke is far less the anguished agnostic than Stapledon. In his widely anthologized story 'The Star' Clarke depicts the spiritual crisis of a Jesuit astrophysicist who discovers the astronomical event that underlay the appearance of the Star of Bethlehem. The realization that an entire peaceful planetary civilization was cremated by its sun going into the nova state shatters the Jesuit's faith. In what sounds like an echo of Stapledon's Eighteenth Men questioning how cosmic beauty can be equated with suffering, he asks, 'But to be destroyed so completely in the full flower of its achievement, leaving no survivors – how could that be reconciled with the mercy of God?'[27] Because the atheist colleagues who share his scientific mission do not ascribe intentionality to Nature none of them suffers in quite the same way as the priest. Believing, with the Psalmist, that the heavens declare the handiwork of God, the Jesuit asks a concluding question that takes on the character of a protest. 'Yet, oh God, there were so many stars you could have used. What was the need to give these people to the fire, that the symbol of their passing might shine above Bethlehem?'[28] For his colleagues, the nova is an act of God in the insurance agent's sense, a deeply regrettable calamity to which no causality can be assigned, but for the Jesuit it is the act of a God unworthy of love or worship. For a man who believes in a deity and whose religious order bears the motto *Ad Majorem Dei Gloria* ['For the Greater Glory of God'] this is the story not of the death of God but of the exposure of God through his acts.

One of the great embodiments of this protest against acts of God occurs in what is arguably the finest apocalyptic film we have, Ingmar Bergman's *The Seventh Seal*. The spokesperson for Bergman in the film is surely not the saintly and philosophical Knight – who is reminiscent in some ways of Mary Shelley's fatalistic Verney – but the Squire Jons, whose crude but humane empathy recalls the various literary incarnations of Noah's wife. During the scene in which a young girl, scapegoated as a witch responsible for the plague, is being burnt it is Jons who bitterly asks the Knight, 'Who watches over that child? Is it the angels, or God, or the Devil, or only the emptiness?' For Jons the child sacrificed to the emptiness becomes a token of the necessity of atheism. If this killing is an act of God, the God in whose name it is authorized is not worthy of belief. The horror of the child's death mirrors a larger, a more universal, an unbearable horror: 'We stand powerless, our arms hanging at our sides, because we see what she sees, and our terror and hers are the same. That poor little child. I can't stand it, I can't stand it ... '[29]

Apocalypse is as deeply interesting a genre of visionary fiction as it has ever been, but 'I can't stand it' has increasingly become the response to the theological implications and the human cost of removing the veil from God's mysteries. There have been occasions when, and there are still groups for whom, apocalyptic disaster is welcomed as an emblem of God's providential care – although this has usually been true only when disaster seemed conveniently reserved for those whom one hated or felt deserving of punishment. Boccaccio could accept the 1348 visitation of the Black Death on the city of Florence as being 'sent down upon mankind for our correction by the just wrath of God.' He could also accept it as a mere 'brief annoyance' that would serve as prologue to the risqué tales told by his aristocratic narrators, who had the means to escape God's corrective action and needed a pastime to fill up the boredom of quarantine.[30] The famous earthquake of 1692 that destroyed and submerged the town of Port Royal in Jamaica was widely celebrated by moralists as a just judgment on a modern Sodom, the act of a righteously offended God. This celebration took place of course in Europe, conveniently distanced from the horrors, where one could imagine that the only victims of the disaster were pirates and prostitutes. More recently, some religious denominations have made similar pronouncements about AIDS, designating it 'the liberals' leprosy' and 'God's curse on those who have perverted scripture.'[31] The spread of the disease to larger and very diverse segments of the population has, however, already made any claim that God devised a remarkably selective plague difficult to maintain even for the most determined fanatic.

In the century of the Somme, the Holocaust of the Jews, the nuclear bomb, contaminated natural resources, a depleted ozone layer, and the indiscriminate terrorist act, the notion of an apocalyptic act of God with moral content has been rendered either ludicrous or appalling. 'Acts of God' are now less likely to draw attention to the justness of the act than to the problem of the god.

Notes

1. H. G. Wells, *A Critical Edition of 'The War of the Worlds'*, ed. David Y. Hughes and Harry M. Geduld (Bloomington: Indiana University Press, 1993), p. 103.
2. William H. McNeill, *Plagues and Peoples* (New York: Anchor, 1976), p. 162.
3. Voltaire's response to Lisbon, in both *Candide* and his *Poeme sur le désastre de Lisbonne*, is discussed usefully in Otto Friedrich's *The End of the World: A History* (New York: Coward, McCann & Geoghegan, 1982), pp. 200–205. See Karen Gershon's poem 'Experiments with God' in *Holocaust Poetry*, ed. Hilda Schiff (New York: St. Martin's, 1995), p. 188.

4. Quoted by Colin Nickerson, 'A World Shaken,' *Boston Globe* (30 November 1997), p. A28.
5. Wells, p. 104.
6. See Hermann Loimer, 'Accidents and Acts of God: A History of the Terms,' *American Journal of Public Health* 86 (January 1996), 101–07.
7. 'Noah's Deluge' in *The Chester Mystery Plays*, adapted into Modern English by Maurice Hussey (New York: Theatre Arts Books, 1957), p. 31.
8. Timothy Findley, *Not Wanted on the Voyage* (New York: Delacorte, 1984), p. 170.
9. John Calvin Batchelor, *The Birth of the People's Republic of Antarctica* (New York: Dial, 1983), p. 107.
10. A persuasive challenge to the conventional reading of *Canticle* as an endorsement of Catholic doctrine can be found in David Seed's 'Recycling the Texts of the Culture: Walter M. Miller's *A Canticle for Leibowitz*,' *Extrapolation* 37 (1996), 257–71.
11. Walter M. Miller, Jr., *A Canticle for Leibowitz* (1959; rpt. New York: Bantam, 1976), p. 240.
12. Miller, p. 292.
13. Miller, p. 301.
14. Mary Shelley, *The Last Man*, 1826 (rpt. New York: Bantam, 1994), pp. 322, 323. See W. Warren Wagar, *Terminal Visions: The Literature of Last Things* (Bloomington: Indiana University Press, 1982), p. 13.
15. The Last Man, p. 423.
16. Olaf Stapledon, *Last and First Men* (London: Methuen, 1930), p. 339.
17. Ibid., p. 6.
18. Ibid., p. 341.
19. Ibid., p. 352.
20. Olaf Stapledon, Preface to *Star Maker* (London: Methuen, 1937), p. vii.
21. Stapledon, *Star Maker*, p. 197.
22. Ibid., pp. 274, 275.
23. Ibid., p. 307.
24. Ibid., p. 283.
25. Ibid., p. 325–26.
26. Batchelor, p. 385.
27. Arthur C. Clarke, 'The Star' (1955), rpt. *Fantastic Worlds: Myths, Tales, and Stories*, ed. Eric S. Rabkin (New York: Oxford University Press, 1979), p. 390.
28. Clarke, p. 391.
29. Ingmar Bergman, Film Script of *The Seventh Seal*, trans. Lars Malmström and David Kushner (London, Lorrimer, 1968), p. 69.
30. Boccaccio, *The Decameron*, ed. and trans. Mark Musa and Peter Bondanella (NewYork: Norton, 1977), p. 7.
31. Susan Palmer, *AIDS as an Apocalyptic Metaphor in North America* (Toronto: University of Toronto Press, 1997), p. 26. See also Robert Marx, *Pirate Port: The Story of the Sunken City of Port Royal* (London: Pelham, 1968).

7

The Dawn of the Atomic Age

DAVID SEED

In the issue of *Life* magazine devoted to the Hiroshima bombing a feature opened: 'Aug. 5, 1945 is the day men formally began a new epoch in their history'.[1] The historical positioning of the bombing of Hiroshima and Nagasaki remains a matter of controversy over whether they marked the end of one period or the beginning of another. Were they the unavoidable necessary final step to destroy the Japanese war machine or an object lesson to the Soviet Union in a superpower confrontation which was already taking shape? Whichever view commentators took there was common agreement that the explosion of the atomic bomb marked a radical turning-point in warfare, and Paul Boyer has shown in his classic study of the bomb's impact on the American imagination *By the Bomb's Early Light* (1985) that, even though there might have been popular acceptance of the bombing, the event was very quickly transposed on to the American scene in a whole series of accounts of atomic attack.[2] In what follows I shall be examining the procedures followed by the two most widely read early commentators on the atomic bomb, William L. Laurence and John Hersey, to see how they articulate their respective convictions that it constituted a turning-point in the natural and political order so radical as to be apocalyptic. Although both writers were journalists they approached the event from the opposite perspectives of producer and victim, and not surprisingly drew on equally opposing conventions of representation.

As soon as Roosevelt decided to press ahead with the Manhattan Project a total clampdown was imposed on all information relating to it. At the same time, however, the director of security General Leslie R. Groves invited the *New York Times* journalist William L. Laurence to become the Project's official chronicler. Part of Laurence's brief was to prepare alternative reports of the Alamagordo test and the Hiroshima bombing in advance of the events, the latter being used by Truman and Stimson, his Secretary of War. In addition he had to prepare a series of articles on the project which would be released to the public after Hiroshima. Laurence was thus at one and the same time given a unique journalistic opportunity, put in a position where he could hardly afford to be critical of the enterprise, and because of security restrictions had time, as Ken Cooper argues, to 'formulate a mythology of the atomic bomb'.[3] It is no surprise then that when he visited the Hanford nuclear reactors in Washington State he depicted them as timeless 'Promethean structures, which may well stand as eternal monuments to the spirit of man challenging nature, [where] mighty cosmic forces are at work such as had never been let loose on this planet in the 3,000 million years of the earth's being'.[4] Laurence abstracts the installations from a specific historical and military context and eternalizes them in a grandiose epic of humanity's drive to uncover the secrets of Nature. He casts himself as an apocalyptic witness granted a privileged glimpse of this massive and silent factory complex. But of course the historical context is only temporarily suspended, and Laurence's ringing rhetoric none too implicitly endorses a national undertaking by replacing the lone mythical hero with the USA. Similarly when Laurence turns his attentions to nuclear bombs his exploitation of the apocalyptic paradigm surcharges the patriotic implications of his accounts.

Laurence's description of the Alamagordo explosion frames itself within biblical accounts of creation. The countdown is delivered by a 'voice ringing through the darkness, sounding as though it had come from above the clouds'. Laurence blanks out the enormous technological enterprise of the Manhattan Project to depict the event as if the observers are witnessing a cataclysm of nature:

> And just at that instant there rose from the bowels of the earth a light not of this world, the light of many suns in one. It was a sunrise such as the world had never seen, a great green super-sun climbing in a fraction of a second to a height of more than 8,000 feet, rising ever higher until it touched the clouds, lighting up earth and sky all around with a dazzling luminosity.
>
> Up it went, a great ball of fire about a mile in diameter, changing colours as it kept shooting upward [...] It was as though the earth had opened and the skies had split. One felt as though one were present at the moment of creation when God said: 'Let there be light.'

> A great cloud rose from the ground and followed the trail of the great sun. At first it was a giant column, which soon took the shape of a supramundane mushroom. For a fleeting instant it took the form of the Statue of Liberty magnified many times.[5]

The dawn is transformed into an apocalyptic spectacle of liberation poised ambiguously between beginnings and endings, birth and destruction. The event is described as supranatural, a sudden manifestation of the primal element of light, and the scientific concept of fission is extended into a rupture of Nature itself. The Manhattan Project scientists are assimilated into a quasi-divine rerun of the act of creation, and through the use of a national icon the dawn of the Atomic Age comes to signify a dawn of American political supremacy; for the explosion, after all, manifests power, the operation of force. Early comments on the atom bomb reflected an ambivalence: for instance, the editor of the victory issue of *Time* magazine reflected ominously that many might be celebrating but 'in the dark depths of their minds and hearts, huge forms moved and silently arranged themselves: Titans arranging out of the chaos an age in which victory was already only the shout of a child in the street'.[6] Truman wondered if it represented the 'fire of destruction' prophesied in the Bible, but in his public announcement after the Hiroshima bombing declared that it represented a 'harnessing of the basic power of the universe'. Similarly press reports expressed satisfaction but fear for the future: 'Yesterday we clinched victory in the Pacific, but we sowed the whirlwind', 'we are dealing with an invention that could overwhelm civilization' and so on.[7] Laurence confines such fears in *Dawn Over Zero* mainly to the countervoice of another observer who sees the explosion as the 'nearest thing to doomsday that one could possibly imagine', in other words as the ultimate ending, not a new beginning.

Laurence strengthened a pattern of iconography (fire-ball, mushroom cloud, etc.) which recurs through later accounts of bomb blasts and in the process highlighted a problem of representation: how to articulate distinctly a process which happens in a millionth of a second and how to capture the sheer scale of the event.[8] Laurence had to rely on the eyewitness accounts of others for Hiroshima. One crew member of the *Enola Gay* stressed the impression of boiling cloud as the mushroom took shape, while another recorded orally the moment-by-moment proliferation of fires and compared the mushroom to a 'mass of bubbling molasses'.[9]

Laurence did, however, manage to see the bombing of Nagasaki for himself, and focuses his description on the metaphorical transformations of a 'pillar of purple fire'.

> Awestruck, we watched it shoot upward like a meteor coming from the earth instead of from outer space, becoming ever more alive as it climbed skyward through the white clouds. It was no longer smoke, or dust, or even a cloud of fire. It was a living thing, a new species of being, born right before our eyes.
>
> At one stage of its evolution, covering millions of years in terms of seconds, the entity assumed the form of a giant square totem pole, with its base about three miles long, tapering off to about a mile at the top. Its bottom was brown, its center amber, its top white.
>
> Then, just when it appeared as though the thing had settled down into a state of permanence, there came shooting out of the top a giant mushroom that increased the height of the pillar to a total of 45,000 feet.
>
> The mushroom top was even more alive than the pillar, seething and boiling in a white fury of creamy foam, sizzling upward and then descending earthward, a thousand geysers rolled into one.
>
> It kept struggling in an elemental fury, like a creature in the act of breaking the bonds that held it down. In a few seconds it had freed itself from its gigantic stem and floated upwards with tremendous speed, its momentum carrying it into the stratosphere to a height of about sixty thousand feet. [...]
>
> The quivering top of the pillar protruded to a great height through the white clouds, giving the appearance of a monstrous prehistoric creature with a ruff around its neck, a fleecy ruff extending in all directions, as far as the eye could see.[10]

Unlike the crewman of the *Enola Gay,* Laurence makes no mention of the city being destroyed, although this would have been an exclusion imposed by the authorities, and concentrates instead entirely on an implicitly sexual process of orgasm and birthing of a titanic monster.

We have to remember that before nuclear tests were started in Nevada and Utah this sort of sublime spectacle was only available to Laurence's readers through such mediations as the passage above. Unlike Wordsworth's description of the Alps in *The Prelude,* readers could not go and experience a nuclear blast for themselves. On the contrary, they were totally reliant on accounts carefully packaged by the security authorities which excluded technical information as well as details of casualties. Laurence temporarily had a virtual monopoly of one of the biggest news stories of modern times which was syndicated in *Time* and *Life* and won a Pulitzer Prize for his efforts. The articles collected in *Dawn Over Zero* (1946) and *Men and Atoms* (1961) position Alamagordo, Hiroshima and Nagasaki within a grand narrative where international scientific endeavour harmonizes with national destiny. Perhaps swayed by viewing events from the perspective of the producers, Laurence had few reservations about nuclear weaponry and was undismayed by the testing of the hydrogen bomb in Bikini Atoll. He had no difficulty in transforming the bigger and better mushroom cloud into a 'protective umbrella that will forever shield mankind everywhere against the threat of annihilation in any atomic war'. Apocalypse for him signified the beginning of a new era, a Pax Americana, and he concluded his 1951 study *The Hell Bomb* by denouncing unilateral disarmament and expressing his conviction that 'good will prevail [...] over the forces of evil [...] that the four freedoms will triumph over the Four Horsemen of the Apocalypse'.[11]

When the fourth bomb was exploded at Bikini Atoll one reporter tried to recapture the sublime tone of previous descriptions, but with mixed results: 'The feature attraction of Operation Crossroads was the atomic bomb cloud. As at Alamagordo, Hiroshima and Nagasaki, this was an immense, luminescent pillar which sprouted majestically after the bomb's first flash. In nine minutes it had climbed to 24,000 feet. [...] The height of the cloud was disappointing.'[12] The phrase 'feature attraction' is the clearest possible indicator that the bomb blast has become a media display divorced from military necessity. Although the other main reporter on Hiroshima, John Hersey, appeared in the pages of *Life* like Laurence, we shall see that he demonstrates a far more sceptical awareness of media processes in his writing.

In the early months of 1946 Hersey was on assignment in northern China reporting on the American marine bases there and on how peace was affecting rural society. Then he got his opportunity, the chance to produce what was originally planned to be a four-part report on Hiroshima which in the event took up the entire *New Yorker* for 31 August 1946, went into book form and became an immediate best-seller.[13] Before turning to that work we need, however, to consider an intermediary text which Hersey read and used. Several months before Hersey's *Hiroshima* there appeared in the *Saturday Review* an article by the Jesuit John A. Siemes entitled 'Hiroshima: Eye-Witness' which superficially resembles the narrative of a single individual in Hersey's book. Father Siemes's account anticipated the broad sequence of Hersey's study in starting with the Japanese expectation of a special weapon on Hiroshima, an hour-by-hour description of the explosion and rescue work, and a conclusion noting the surprising lack of anti-American feeling and speculating mildly on the ethics of the superbomb. Father Siemes evidently intended to capture the immediacy of the event by writing most of his account in the present tense, but the effect was rather different. Consider, for example, his description of the explosion:

> Suddenly – the time is approximately 8:15 – the whole valley is filled by a garish light which resembles the magnesium light used in photography, and I am conscious of a wave of heat. I jump to the window to find out the cause of this remarkable phenomenon, but I see nothing more than that brilliant yellow light. As I make for the door, it doesn't occur to me that the light might have something to do with enemy planes. On the way from the window, I hear a moderately loud explosion which seems to come from a distance and, at the same time, the windows are broken in with a loud crash. There has been an interval of perhaps ten seconds since the flash of light. I am sprayed by fragments of glass. The entire window frame has been forced into the room. I realize now that a bomb has burst and I am under the impression that it exploded directly over our house or in the immediate vicinity.[14]

Siemes works on the mistaken assumption that the mere statement of impressions is enough to guarantee immediacy, but the latter is muffled by his desire to convey information (time and resemblance) in the opening sentence. Siemes has no capacity to capture the pace of sense impressions which must have run well ahead of any witness's capacity to draw inferences, giving an effect of running commentary on his own actions as if watching himself at every second. More damagingly, the impression of instantaneity was constantly being belied by details which must only have become clear after the event.

A progression takes place in Siemes's account from personal glimpses (present tense) to the assembly of an overall narration (past tense). This progression is traced out as a physical journey from the Catholic Novitiate (some two miles outside Hiroshima) towards the centre of the city. As the day after the blast dawns he sees before him a scene of utter devastation: 'Where the city stood, everything, as far as the eye can reach, is a waste of ashes and ruin. Only several skeletons of buildings remain. The banks of the river are covered with dead and wounded, and the rising waters have here and there covered some of the corpses.'[15] The sublime effects of the bomb blast described by Laurence and others depend on a vertical expansion which dwarfs the observer. Here, by contrast, Siemes gives us a panorama which expands the field of vision laterally towards an unidentifiable horizon. The impact now is not awe but a dislocation arising from the eye's incapacity to see limits to this expanse. And far from being pure spectacle, buildings and inhabitants collapse together in a common reduction to ruin and inert matter. This visual moment is one of the most powerful in an account which had an impact on Hersey. When he arrived in Hiroshima he first met Father Kleinsorge, Siemes's Father Superior, who then introduced him to his other witnesses. Hersey took a more sympathetic view than Siemes on the lack of a co-ordinated rescue strategy by the Japanese and towards the end of *Hiroshima* quotes from Siemes's conclusion.

In addition to drawing on Siemes, *Hiroshima* was partly modelled on Thornton Wilder's 1927 novel *The Bridge of San Luis Rey*, which Hersey read during a bout of illness on a U.S. destroyer and which uses the disaster of a bridge over a gorge in Peru collapsing to examine the meaning of the catastrophe.[16] A Catholic cleric, one Brother Juniper, sets out to understand the lives of five victims because of their metaphysical implications: 'If there was any plan in the universe at all, if there were any pattern in a human life, surely it could be discovered mysteriously latent in those lives so suddenly cut off. Either we live by accident and die by accident, or we live by plan and die by plan.' Only two starkly opposing possibilities are posited, either random chance or divinely directed design: 'Some say that we shall never know, and that to the gods we are like the flies that the boys kill on a summer day, and some say, on the contrary, that the very sparrows do not lose a feather that has not been brushed away by the finger of God'.[17] Wilder assembles three consecutive narratives of the victims' lives which all converge on the final moment of the disaster but, although Brother Juniper assembles an enormous volume to demonstrate the workings of destiny, the book is burnt by the authorities and its author accused of heresy. The demonstration of logical causality therefore remains an elusive dream within the novel. Hersey modifies Wilder's method in a number of important respects since turned into cliché by disaster narratives. Firstly there is no Brother Juniper; his role is implicitly woven into the narrative voice. Secondly Hersey deals with survivors not fatalities. And thirdly he counterpoints these characters' destinies against each other in a form of montage.

By 1945 Hersey had established himself for journalistic reports on the war in journals like *Life* and *Time* many of which were published in novel form. *Into the Valley* (1943), for example, describes a marine battle on Guadalcanal and explicitly uses reportage to help with the war effort. If readers could know about the combat experience, he explained, 'I think we would be an inch or two closer to winning the war and trying like hell to make peace permanent.'[18] By 1946 of course the larger war narrative which Hersey could earlier take for granted had concluded and a new era had come into being. William Laurence made no bones about witnessing a radical historical change, starting his essay 'The Atomic Age Begins' as follows: 'I watched the birth of the atomic age from the slope of a hill in the desert land of New Mexico.' Spencer Weart has shown that the metaphor of the nuclear bomb as a baby 'delivered' from its maternal carrier by male technology very quickly became institutionalized after Hiroshima.[19] And similarly John Siemes's account of the bombing was already packaged in the *Saturday Review* as an 'Atomic Age' byline, complete with the famous atomic nucleus logo and punctuated by triumphalist graphics celebrating America's military might.

Hersey made every effort to avoid such triumphalism by focusing on the plight of victims, five Japanese and one German Jesuit. None of the characters appears to have made any contribution to military activity; instead they are described as civilians with their own domestic and professional worries which place them at the farthest extreme from the cartoon fanatic. Indeed one character, a Methodist pastor, has studied in the USA. In countless ways Hersey familiarizes these characters to his western readers, only interrupting his narratives to point out instances of cultural difference like the

Japanese treatment of their dead or their concept of shame. Part of the book's polemic consists of the steady accumulation of details which force the reader to situate him/herself imaginatively the victims' daily lives. The narrative method he uses is strategic. For David Sanders 'Hersey's account ... is strictly limited to the "visual horizon" of his six survivors; and it so skilfully renders the details of that horizon as to establish suspense for even his best informed readers.'[20] For 'visual' we should read 'perceptual' since Hersey does a thorough job of conveying the sensory immediacy of his characters' experiences.

The first chapter of *Hiroshima* compresses the narrative sequence traced out at length by Thornton Wilder for the catastrophe in *The Bridge of San Luis Rey* functions teleologically as an end-point. For Hersey's six survivors, however, the bomb blast occurs within a continuing sequence as we shall see. Here, for instance, is how Hersey describes the experience of Miss Toshiko Sasaki, a clerk in the East Asia Tin Works:

> Everything fell, and Miss Sasaki lost consciousness. The ceiling dropped suddenly and the wooden floor above collapsed in splinters and the people up there came down and the roof above them gave way; but principally and first of all, the bookcases right behind her swooped forward and the contents threw her down, with her left leg horribly twisted and breaking underneath her. There, in the tin factory, in the first moment of the atomic age, a human being was crushed by books.[21]

Superficially this passage gives an impression of stylistic naivete, of paratactic syntax where each event is given the same priority. As in early Hemingway, however, the style masks its own sophistication. The two most frequent verbs in *Hiroshima* are not surprisingly 'fall' and 'throw'. Structures and humans alike collapse and individuals are literally and metaphorically thrown out of their routine lives into a predicament of common need. Here a series of falls occur within which Hersey, not Miss Sasaki, readjusts the chronology and prioritizes that of the bookcases because of their symbolic potential. Paul Boyer has shrewdly remarked that Hersey quietly recognizes the problematic nature of style in this situation by consistently referring to books as an irrelevance.[22] Indeed if books are taken as a privileged sign of civilization then Hersey renders the clerk's fate as an ironic parabolic image of humanity being destroyed by its own culture.

Hersey himself later explained that he had deliberately avoided moralizing or preaching, preferring to 'help readers to find their own deepest feelings about this new instrument of killing', and had chosen a 'flat style' to avoid giving the reader a sense of mediation.[23] But then we have already seen how the style covertly guides the reader to take up positions with regard to the bombing. It is this veiled appropriation of his witnesses by Hersey that Alan Nadel has cogently criticized. Firstly he declares that Hersey is claiming more narrative authority than he is entitled to; then he blanks out the mediation of the witnesses' accounts through interview and recollection; and finally Hersey gives no clues about the process of editorial selection which took place in the composition of his account.[24] *Hiroshima* opens with a collective query: how could these six have survived? Here another connection with Wilder emerges. Brother Juniper is haunted in compiling his narrative by the 'fear that in omitting the slightest detail he might lose some guiding hint' and as a result attempts a ludicrously encyclopedic account.[25] Hersey's characters are similarly bemused by the 'small items of chance or volition' which might explain their own survival. Hence Hersey's extraordinary care to specify their exact positions at the moment of the bomb blast because a few inches nearer or farther from a window, for example, could literally make the difference between life and death. Dwight Macdonald totally misunderstood the work when he objected that Hersey had 'no eye for the one detail that imaginatively creates a whole', because a whole is exactly what the survivors cannot glimpse.[26] The bomb blast is now apocalyptic not as visual spectacle but in bringing about a sudden and total rupture in their lives. One character's disorientation is figured as physical movement: he is 'thrown forward and around and over'. Another character finds his room reduced to 'weird and illogical confusion'.[27] *Hiroshima* thus describes a drama of bewilderment where characters are suddenly cut off from relatives, possessions and habitations, and forced as a result to search for what is lost.

Although Hersey divides his book into approximate phases these are chiefly characterized by their narrative *dis*continuity. Here are Miss Sasaki's experiences once again:

> Much later, several men came and dragged Miss Sasaki out. Her left leg was not severed, but it was badly broken and cut and it

> hung askew below the knee. They took her out into a courtyard. It was raining. She sat on the ground in the rain. When the downpour increased, someone directed all the wounded people to take cover in the factory's air-raid shelters. 'Come along,' a torn-up woman said to her. 'You can hop.' But Miss Sasaki could not move, and she just waited in the rain. Then a man propped up a large sheet of corrugated iron as a kind of lean-to and took her in his arms and carried her to it. She was grateful until he brought two horribly wounded people – a woman with a whole breast sheared off and a man whose face was all raw from a burn – to share the simple shed with her. No one came back. The rain cleared and the cloudy afternoon was hot; before nightfall the three grotesques under the slanting piece of twisted iron began to smell quite bad.[28]

In these lines Hersey brilliantly captures the isolated dependency of a war casualty. Rendered helpless and immobile, Miss Sasaki must wait for help which is rendered by a series of unnamed figures who themselves – like the 'torn-up woman' – might be casualties. This account becomes a study of absolute contingency where help might or might not come. The context of the war is not quite forgotten but skewed like most details in the book so that an air-raid shelter (now obsolete) can only help the victims to get out of the rain. Duration is measured out in terms of arrivals without which Miss Sasaki joins her fellow victims as a helpless and debilitated organism waiting for death. *Hiroshima* then foregrounds discontinuous images in order to convey the rupturing of perceptions of reality brought about by the bomb. Expectations are always reversed. A shelter turns out not to have offered protection against wounding and casualties' features become transformed: eyes into tears, mouths into wounds, and so on.

The survivors' experience is disabled twice over by their separation from each other and from the virtual inaccessibility of the news media. At the time of the bomb expectations were rife of an imminent bombing or even invasion by the Americans. Accordingly the survivors latched on to single characteristics of the atom bomb in order to make sense of their experience. The scale of the blast was put down to self-scattering bombs; the smell of ionized air to a gas attack; and the flash to magnesium powder (the last of these being dreamt up by a newspaper reporter). The absence of the media is not presented as a simple lack of information, rather as another example of an absurd discontinuity between official statements and the survivors' experience: 'even if they had known the truth, most of them would have been too busy or too weary or too badly hurt to care that they were the objects of the first great experiment in the use of atomic power, which (as the voices on the short-wave shouted) no country except the United States, with its industrial know-how, its willingness to throw two billion gold dollars into an important wartime gamble, could possibly have developed'.[29] Once again Hersey ironically distances himself from the celebratory descriptions of the beginning of the atomic age as a triumph of American military and industrial know-how. A passage like this cannot be explained by William J. Scheick's complaint that Hersey uses a double voice in *Hiroshima*: 'There seems to be [...] two contrary vectors in the same narrative voice in his account: the existential manner of a dispassionate observer concerned with philosophical insight into humanity and the humanistic manner of a childlike victim dazed in reaction to a traumatizing concrete situation.' For Scheick this is a sign of the postwar period because the latter aspect 'at the same time represents the anomie and loss of affect that are Cold-War phenomena'.[30] In a sense Scheick is complaining about a rhetorical doubleness in Hersey's use of the third person which enables him to move in and out of his survivors' consciousnesses, sometimes using them as temporary angles of vision, at other points describing them as if seen by another. It would be inconceivable for there *not* to be some doubleness in Hersey's narrative voice, since he was attempting the difficult task of describing the experiences of a cultural group until very recently thought of as the enemy for members of the very nation which manufactured and dropped the bomb. Hersey manages this delicate balancing act by distancing himself from the occupying U.S. authorities, drily noting that far more information on the atom bomb was circulating in Japan than in the USA.

Etymologically 'apocalypse' denotes an uncovering or revelation, which helps to explain why, for all their differences, Laurence and Hersey are profoundly visual writers, although each concentrates on different spatial areas. Laurence scans the skies while Hersey keeps his perspectives fixed at ground level. For Laurence what is revealed is a new entity, a new embodiment of force; while Hersey sees apocalypse in the radical disruption of civilian life. When one of his witnesses tells another 'things don't matter any more' he is identifying a break-down in the system of commodities which is giving way to greater and more pressing human need. Both Laurence and Hersey present the Bomb as a sudden rupture of the natural order. American witnesses to the bomb blast all agreed, for instance, that the flash was like a super-sun. Hersey finds a more complex set of features which do not rely only on magnitude. Now destruction seems freakishly inconsistent; dress patterns are transferred on to the bodies of survivors; the outlines of those near ground zero imprinted on walls, and the sudden growth of plants in the ruins. Like Siemes, Hersey works towards a lateral panorama of Hiroshima which suggests in still life a complex continuity rather than a total ending. The human dwellings have been the most thoroughly destroyed ('range on range of collapsed city blocks') and they establish a background for the surviving structures:

> [...] naked trees and canted telephone poles; the few standing, gutted buildings only accentuating the horizontality of everything else (the Museum of Science and Industry, with its dome stripped to its steel frame, as if for an autopsy; the modern Chamber of Commerce Building, its tower as cold, rigid, and unassailable after the blow as before; the huge, low-lying, camouflaged city hall; the row of dowdy banks, caricaturing a shaken economic system); and in the streets a macabre traffic – hundreds of crumpled bicycles, shells of streetcars and automobiles, all halted in mid-motion.[31]

This image occurs in the book immediately after Japan's surrender and totally avoids triumphalism by being focalized through a German Jesuit, i.e. through the eyes of one outside the economic and political system of Japan. Hersey implies that the only winner is a commercial structure (and a financial one; the banks are 'shaken' but essentially intact) indifferent to the human casualties. The 'traffic' in the streets mimics city life as if frozen in a snapshot but is rendered grotesque by the total absence of human beings. Once again we have a tensed combination of life and death. The scene itself denotes both destruction and survival; but even more importantly, it is focalized through a survivor coming to terms with a cataclysmic event.

Here we can locate the key differences between Hersey and Laurence. The latter mythologizes a finite event through rhetoric and imagery which uncritically support America's triumphalist pride in her military superiority at the beginning of the Cold War. Hersey on the other hand stresses how Hiroshima ruptured the lives of his chosen survivors. In 1985 he added a sequel to his original account called 'Aftermath' which does not generalize the implications of his original study but which modifies it in two important aspects. Early official reports on Hiroshima and Nagasaki, for obvious reasons, attempted to minimize the damage caused by radiation, but Hersey takes Father Kleinsorge as a case study to demonstrate, as Robert Jungk does in his *Children of the Ashes*, that the bombing was a continuing catastrophe. Kleinsorge endured for the rest of his life a relentless series of painful ailments which constantly baffled the doctors by their seeming illogicality, and only found relief in death. Secondly, Hersey uses the pastor Tanimoto to demonstrate the ironies in media processes. Because MacArthur had forbidden the release of information on the two bombings within Japan Tanimoto could only campaign about their effects when abroad, specifically when in the USA. There he was adopted as a cause by Norman Cousins, the editor of the *Saturday Review* and on 11 May 1955 the hapless Tanimoto found himself transformed into a media personality on the NBC programme 'This is Your Life' which opened with predictable shots of the rising mushroom cloud. While Hersey's strategy of minimizing overt narrative comment in *Hiroshima* cannot be seen as a strategy whereby he totally avoids the appropriation of his survivors for narrative purposes, nevertheless his characters' priorities are allowed to stand in the foreground.

Notes

1. 'The Atomic Bomb', *Life* 19.viii (8 August 1945) p. 87B.
2. Paul Boyer, *By the Bomb's Early Light* (2nd edition Chapel Hill and London: University of North Carolina Press, 1994), p. 5. One of the first such accounts was *The Murder of the U.S.A.* by Murray Leinster writing as Will F. Jenkins, which describes the destruction of one-third of the nation in only forty minutes.
3. Ken Cooper, 'The Whiteness of the Bomb', in *Postmodern Apocalypse: Theory and Cultural Practice at the End* (Philadelphia: University of Pennsylvania Press, 1995), p. 105.
4. William L. Laurence, 'My Life in Atomland', *Men and Atoms* (London: Scientific Book Club, 1961), p. 98.
5. Laurence, 'The Atomic Age Begins', in *Dawn Over Zero: The Story of the Atomic Bomb* (1946; London: Museum Press, 1947), pp. 9–10. Laurence revised this description extensively for his collection *Men and Atoms* (1961).
6. Editorial, *Time* (20 August 1945), p. 19.
7. David McCullough, *Truman* (London and New York: Simon and Schuster, 1993), pp. 443, 455, 456. In public Truman was more upbeat announcing that the bomb was a 'harnessing of the basic power of the universe' and declaring America's custodial responsibilities: 'we must

constitute ourselves trustees of this new force – to prevent its misuse', *Truman*, p. 455; 'Atomic Age', *Time* (20 August 1945), p. 29.

8. General Farrell, one of the witnesses to the Alamagordo test declared frankly: 'words are inadequate tools for the job of acquainting those not present with the physical, mental, and psychological effects. It had to be witnessed to be realized.' Robert Jungk, *Brighter than a Thousand Suns* (Harmondsworth: Penguin, 1964), p. 184.
9. Gordon Thomas and Max Morgan-Witts, *Ruin from the Air: The Atomic Mission to Hiroshima* (London: Sphere, 1978), p. 431.
10. Laurence, '12:01 over Nagasaki', *Men and Atoms*, pp. 159–60.
11. Laurence, *Men and Atoms*, p. 197; *The Hell Bomb* (London: Hollis & Carter, 1951), p. 113. The Four Freedoms were enunciated by Roosevelt in his annual message to Congress on 6 January 1941 and consisted of freedom of expression and of worship, and freedom from want and from fear.
12. 'Bikini's Atomic Blast', *Life* 21.iii (15 July 1946), p. 26.
13. For information on responses to Hersey's *Hiroshima* v. Michael J. Yavenditti, 'John Hersey and, the American Conscience: The Reception of *Hiroshima*', *Pacific Historical Review*, 43 (1974), pp. 24–49; and Boyer, pp. 204–10.
14. John A. Siemes, 'Hiroshima: Eye-Witness', *Saturday Review* (11 May 1946), p. 24.
15. Ibid., p. 41.
16. Biographical details from Hersey's 1988 interview made for the American Audio Prose Library.
17. Thornton Wilder, *The Bridge of San Luis Rey* (London: Longmans, Green 1931), pp. 5, 8. The last passage half-quotes Gloucester's famous lines from *King Lear*, 'As flies to wanton boys, are we to the gods;/ They kill us for their sport' (IV.i).
18. John Hersey, *Into the Valley* (New York: Knopf, 1943), p. 4; Boyer, p. 204.
19. Laurence, *Men and Atoms*, p. 115; Spencer R. Weart, *Nuclear Fear: A History of Images* (Cambridge, Mass.: Harvard University Press, 1988), pp. 147–8.
20. David Sanders, *John Hersey* (New York: Twayne, 1967), p. 46.
21. John Hersey, *Hiroshima* (Harmondsworth: Penguin, 1986), p. 23.
22. Boyer, p. 208.
23. Yavenditti, p. 35; Boyer, p. 208.
24. Alan Nadel, *Containment Culture: American Narratives, Postmodernism, and the Atomic Age* (Durham, N C : Duke University Press, 1995), pp. 55–7.
25. Wilder, p. 129.
26. Quoted in Boyer, p. 206.
27. *Hiroshima*, pp. 15, 30.
28. *Ibid.*, pp. 44–5. Some details from this passage bear a striking similarity to Hemingway's description of political executions in the vignette to Chapter 5 of *In Our Time*.
29. *Ibid.*, p. 66.
30. William J. Scheick, 'The Binuclear Voice of Hersey's *Hiroshima*,' *Nuclear Texts and Contexts*, 11 (Spring 1995), p. 5.
31. *Hiroshima*, p. 88.

8

Silo Psychosis: Diagnosing America's Nuclear Anxieties Through Narrative Imagery

CHARLES E. GANNON

The Cold War psychology – and often, psychopathology – of American culture was frequently acted out, and tellingly disclosed, through narrative articulations of its nuclear stream of consciousness: symbolic shadow-plays of new weapons and maybe-wars with which it both amused and indoctrinated itself over a long period of mushroom-clouded M.A.D.-ness. However, an optimally rigorous and revealing inquiry into American obsessions with, and representations of, the bomb might best be achieved through a comparative analysis which employs cross-cultural contrasts to bring the unique characteristics of U.S. nuclear nightmares into high relief. In particular, comparisons between British and American images of the bomb – and its aftermath – can be used to highlight the key features of America's nuclear psychology, to explain their uniqueness, and to reveal the influence of nuclear weapons upon the consciousness of both the nation's political elites and general public.

In British and American narratives, images of the actual moment of nuclear destruction are notable mostly for their similarities. This is hardly surprising, since the actual conflagration is virtually the only aspect of the nuclear question that is (mostly) congenial to straightforward, objective analysis: the physics is not subject to contextualization. However, in British versus American nuclear war films, the selection of what to frame and how to frame it is subtly different.

In *Threads* (1984), as in other U.K. films which envision the moment of atomic incineration, the portrayal of the blast effects is frank, fast, even brutal.[1] However, in keeping with a tradition dating back to Wells's ur-text of nuclear conflict, *The World Set Free* (1914), the intensity of these visual examinations is leavened by a marked degree of circumspection, by an unwillingness to step back to acquire a macroscopic view of the calamity.[2] Accordingly, in *Threads* – and also in *The Wargame* (1965) and *The Bedsitting Room* (1969) – the audience is never provided with a bird's-eye look at the end of the world. The scale on which we witness the effects of the bomb never exceeds the devastation of individual buildings; the scope never expands to show a larger landscape. As in *The World Set Free* and its British progeny, we certainly see the human effects of the blast, but each view is intimate, immediate. Hence, it might be said that integral to many British portrayals is a marked reluctance to look too closely or too long into the infernal face of the atomic Asmodeus. Therefore, if we were to categorize British cinema's response to nuclear war psychologically as either 'approach' or 'avoidance', it would seem that 'avoidance' is by far the dominant tone.

The exact opposite holds true in the States. Indeed, the American tendency to approach – and reapproach – images of Armageddon is pronounced and unsettling. The most striking aspect of U.S. depictions of the moment of atomic incineration is the strange mix of shocked horror and obsessive compulsion that seems to underlie these representations of larger-than-life destruction. U.S. films, as epitomized by *The Day After* (1983), revel in detailed visual dissections of the detonation and offer expansive views of mass destruction.[3] More disturbing still, this fascination often becomes an obsession in American films, one that reaches its grim zenith in the action/sf film genre. For instance, the nuclear blast sequence from *Terminator 2: Judgment Day* (1991) is horrible, even repulsive, yet the obsessive character of the visuals – the attention to detail, the dilation of time to permit an almost clinical assessment of the annihilating effects of the bomb – strongly suggest an almost erotic fascination with the spectacle of destruction.

However, it would be unfair and inaccurate to suggest that graphic depictions of Armageddon are simply indicative of an American willingness to pander to the sensationalistic and dramatic. As Patrick Mannix points out in his incisive and extensive examination of Nicholas Meyer's *The Day After*, the cinematic scaling and framing of nuclear devastation is at least as much informed by rhetorical objectives as it is by a perverse hunger for lethal spectacles: 'Because Meyer wants to show the effects of nuclear war on a cross-section of people, he must keep cutting from one character to another in the large cast, never concentrating on one long enough to develop any real psychological depth.' This technique is often cited as a weakness of the film, which, according to Mannix, indicates that, instead of education,

> fear seems much more crucial to its rhetorical strategy. Meyer begins by insuring the maximum level of identification for an American audience through his choice of setting: Lawrence, Kansas, almost the exact center of the nation ... (F)armers harvest corn in the nearby fields; children play in a park; workers in a milk plant run their bottling machines; students go to classes at the university ... (T)hese are meant to be glimpses of the lives of people like us – lives that are about to be destroyed by a calamity that threatens us daily. Through this identification, Meyer is conditioning us to feel the sense of imminent danger that fear demands.[4]

Further evidence supports Mannix's contention that Meyer's narrative constructs are designed to generate a growing sense of fear and helplessness. The onset of the nuclear conflict is foreshadowed through an accumulation of half-heard television and radio reports on escalating international tensions, clues of the coming apocalypse that we understand with clarity and certainty, but which many of the characters in the film either ignore or discount. Consequently, we watch them proceed to their doom with the same sense of dread that grows in us as we watch an ingénue in a slasher movie enter a house in which we know the pathological killer is lurking. We are unable to deter this sacrificial starlet, ready to flinch every time she enters a new room, uncertain when the blade will fall, but certain that it will. According to this formula, we are powerless to stop what will transpire, but in *The Day After*, this horror of entrapment expands from several moments to many minutes, the anxiety and inevitability of doom extending into a dreadful, slow-motion anticipation of the multimegaton lacerations that will surely shred the unexpecting social tissue of the body politic itself.

When Meyer finally does let the nuclear knife fall, its slashing is neither brief nor oblique: as Mannix points out, 'Having set up the identification that will provide the groundwork for fear, Meyer attempts to exploit the repertoire of nuclear dangers by showing their effects on these people.' The montage of shots depicting the actual atomic devastation includes the immolation of countless individuals (whose skeletons improbably persist a fraction of a second after they are vaporized); tidal waves of star-hot plasma washing away trees, animals, flesh; the disintegration of buildings; and a sustained collage of nuclear test footage showing the horrific (and often bizarre) effects of the bombs themselves. But this four-minute sequence is just the beginning of Meyer's studied (and some suggest, tiresomely didactic) exercise in viewer education via emotional shock therapy: the strongest doses are to be found in his depiction of the aftermath. As Mannix observes, 'True to the facts of nuclear war – and Meyer is scrupulous in making sure that his depiction is accurate – the survivors of the missile strikes face an ongoing struggle for life.'[5] Indeed, Meyer's extended investigation into the hellish physical, emotional, and social challenges that must be surmounted in order to survive in the aftermath of a nuclear attack is atypical, in that most American films that depict a post-apocalyptic world have been 'bomb-sploitation' adventure romps. However, what does emerge as a constant – both in 'serious' nuclear war films and the countless 'Mad Max' rip-offs – is the fascination with panoramas of annihilation and destruction.

If the British narrative's tendency to avert its gaze from the heart of the nuclear horror is easy to understand, the American narrative's tendency to stare fixedly at this spectacle is not. Susan Sontag offers us one explanatory postulate in her essay 'The Imagination of Disaster'. Speaking only of films, and in particular, their mesmeric fixation upon scenes of destruction, she proposes that they may effect a form of catharsis, a grappling with the unimaginable in order to feel that some measure of control has been asserted over that which cannot be controlled.[6] However, a more pertinent model may be found in Vivian Sobchack's 'The Violent Dance: A Personal Memoir of Death In the Movies'.[7] Sobchack proposes that the avid American interest in the depiction of violence is, in effect, an attempt at psychological prophylaxis against the shock we would suffer should we actually become witnesses to such horrors. In terms of atomic war, this would represent an attempt to prepare for the unimaginable by pre-emptively shocking – and partially inuring – ourselves to the horrors of the ultimate in annihilation.

Whatever the reason – dark prurience, futile catharsis, or a desperate attempt at self-preparation – American viewers seem to possess at least some, if not all, of the mixed motivations that almost certainly compelled Lot's wife to turn and consider the conflagration that consumed Sodom. American film makers and viewers look long, hard, and possibly too deeply into the hypnotic flames, so fascinated and compelled that the normal reaction to the horrific – an impulse to look away – is overridden. In contrast, the behavior of the British viewing public seems more reminiscent of that associated with Noah, who is uncomfortable with the idea of listening too closely to the profound voice, and prescription, of doom. It is almost as if there is an underlying, unstated fear that if the foretelling of the coming flame deluge is heard too clearly, in too much horrible detail, then it may also immolate the consciousness of the listener.

However, the differences in British and American depictions of nuclear warfare become more pronounced – and more revealing – when the narratives move beyond the actual moments of annihilation into the horribly altered world of the atomic aftermath. In films originating in the U.K., the first – and primary – focal image of this aftermath is almost invariably a city – another trend with considerable roots in Wells's original work, as well as in the memories of the Blitz. In films such as *Threads, The Bedsitting Room,* and *1984* (1984), the dominant British image is that of the crushed city: masonry and girders tumbled into ramshackle streets, hollow-eyed survivors picking through the urban graveyard in search of the dead, the dying, and the shreds of yesteryear, which might have ended only a second before. It is a worldview in which the dominant themes are those of desolation and defoliation, which, if considered in the specific context of an urban landscape, may reveal the equally specific fears that underlie this trend in U.K. imagery.

To the British mind, it seems, the destruction of the city – not any particular city, but the sheer concept of 'city' – is the nexus of national and social terror. At one level, this may reflect the centripetal mental and cultural force exerted over Great Britain by the city of London and the revealingly-named 'Home Counties'. Certainly no single city or region in the U.S. has ever been comparably central to the national consciousness, governance, and history. Consequently, it might be suggested that the loss of *a* city is actually a cipher for the loss of *the* city of London – and with it, the loss of nation, culture, identity. This is certainly gestured to in both the earliest British contemplations of apocalypse (in the ruined London of both *The World Set Free* and, particularly, *The War of the Worlds* (1898)) and in its latter-day manifestations.[8] But should we therefore conclude that the destruction of Sheffield in *Threads* is merely a symbol for a similar annihilation that we assume is occurring alongside the Thames?

Such a conclusion would be both ill-advised and dangerously simplistic. Instead, the imagery of urban annihilation may indicate a specific, significant geopolitical reality that informs the British mind, but which is largely absent from American thought. This reality is that of the 'Island Fortress' self-perception – which, in the modern age of push-button apocalypse, undergoes a terrifying inversion: Britain's traditional sense of defensive security becomes an anxiety of horrible, claustrophobic vulnerability. Where once England's relatively small size and marine separation offered it a measure of impregnability, daunting even the onrushing Nazi armies only 55 years ago, the emergence of the bomb has converted the Green and Pleasant Land into a killing zone: a small parcel of land which can be easily blanketed by multiple strikes. In its limited confines, there is no room to run; there are no prairies, outbacks, or mountains so remote that refugees might entertain the hope of escaping the pestilential aftermath or subsequent hordes of the desperate and barbarous. Once the bombs go off, there are no longer any alternatives; there are no escapes, no distant horizons to disappear behind. The people of Britain are trapped – as are the people of a city – between shores and among streets that can only continue to be piled higher with the dead, and after, must crumble down in unarrestable decay.

Conversely, American nuclear war narratives rarely pursue their exploration of aftermath so exclusively in cities. There is, in addition to urban perspectives, extensive consideration of the destruction of the countryside and of isolated towns – whether or not the damage there was inflicted directly by the blast or as 'collateral effects.' Indeed, much of the post-blast action of American narratives (in both text and film) takes place away from urban centers, as is the case in stories as diverse as *The Day After*, *Testament* (1983), Pat Frank's *Alas, Babylon* (1959), Whitley Stieber and James Kunetka's *Warday* (1984), and Walter M. Miller's *A Canticle for Leibowitz* (1959).[9]

This, on reflection, should hardly surprise us. Whereas English nuclear consciousness may be informed by island-claustrophobia and a 'mind-the-gap' appreciation of the dangerously narrow separation between 'life as usual' or 'death among the ruins', the American nuclear consciousness has been influenced by a completely different geopolitical reality, which correspondingly gives rise to completely different attitudes, reactions, outlooks, and a completely different palette of images. For want of a better term, this different geopolitical perspective might be called the frontier mentality: the inveterate American belief that there is always a new land beyond the horizon, a new mountain beyond which to start a new life, an undiscovered country ready to welcome the determined and the industrious. Whether this is infantile ingenuousness or realistic optimism is difficult to say, but in apocalyptic scenarios, the comparative vastness of America often functions subconsciously as a kind of sponge for atomic bombs, allowing the nation to absorb damage.[10] By contrast, the British assessment of post-nuclear survival effectively plummets to zero the moment a missile is launched in anger. This is quite unlike its American counterpart, which describes a more gradual gradient of catastrophe. This belief in the 'survivability' of potentially 'limited' nuclear attacks seems, in turn, to promote a greater willingness to balance the concept of 'the bomb as tool' against the undeniable horrors of its use.

Hence we see an image of the bomb in American literature and film that has no analogue use in the U.K.: the bomb as prophylaxis. The entirety of Europe – so densely populated, so delicately balanced in its interconnections – shares a common awareness that the use of the bomb means the breakdown of everything. Except in extremely remote regions, there is not even enough space to establish a 'dead zone' in which to detonate a nuclear device safely. But in the American consciousness, filled as it is with images of North Dakotan prairie, Texan badlands, and Southwestern deserts (such as that around Alamagordo), there is a pronounced temptation to think in terms of places in which it would be 'safe' to use a bomb.

Regardless of whatever purely clinical accuracy might reside within such an assertion, it propagates a potentially perilous mentality. Individuals come to see the bomb as survivable and, as a result, not so much as an outrage against civilization as a 'tool' – a utility object which, under the right conditions, might be of benefit to the nation, or even the world. And it is through this dangerous doorway of recontextualization that American narratives arrive at the bomb as prophylaxis. Consider the scene from the George Pal adaptation of Wells's *The War of the Worlds* (1953), in which the Martians are introduced to (and are unimpressed by) the bomb, or Larry Niven and Jerry Pournelle's novel *Footfall* (1983), in which a good-sized chunk of the American prairie is sacrificed in order to annihilate an alien landing force. Then there are the numerous B-movie homages to the mushroom cloud's (supposed) potential to either disincline or discorporate threatening monsters/mutants/ maladies as diverse as Godzilla, virally created zombies, and overgrown amoebae. However, as often as not, these attempts at atomic eco-control only serve to further perturb, promote, and/or propagate the threatening agency. And although these narratives are as ludicrous as they are predictable, they none the less lie only one small discursive step away from a notably responsible and somber analog: the bomb is proposed as a means of effecting widespread sterilization over an area infected with Michael Crichton's deadly Andromeda Strain (1969).[11] However, the research scientists learn in the nick of time that this course of action would also have undesirable results: global dissemination of the utterly lethal Andromeda virus.

Such cautionary endings notwithstanding, America's atomic utilitarianism is most noteworthy for the seriousness and influence of its most significant manifestations. The so-appropriately-named M.A.D. (Mutual Assured Destruction) strategy conceived of the bomb as the insurance, rather than antithesis, of peace. In Eugene Burdick and Harvey Wheeler's *Fail-Safe* (1962), the bomb serves theoretically as a means of redistributing international justice and balance (*à la* Hamurabi): the erroneous incineration of Moscow by an American bomber threatens global catastrophe, until America re-equalizes the scales by dropping a similar weapon on New York City. This cultural perspective also helps us grasp the logic underlying the use of the bomb at Hiroshima itself – the use of a nuclear tool, justified by claims that it would ultimately save lives by making Operation Olympic (the planned American invasion of Japan) unnecessary.[12] Despite these alarming examples, the issue here is not to discuss or even highlight the moral aspects of these decisions, but, rather, to call attention to the conceptual heritage in American thought that allows a place for the image of the bomb as prophylaxis. And it may be that this mindset helped give rise – and shape – to what commentators refer to as the American Century, for the relationship between the bomb and American attitudes played a large role in defining the new political position, praxis, and identity that abetted America's evolution into a modern superpower.

This evolution was facilitated not merely by technological attainments, but by alterations to the national psychology. One such psychological alteration involved the increasing perception of other individuals as part of the 'masses'. In military matters, this requires extreme detachment from the mitigating emotions of empathy and compassion, culminating in a systematic and absolute dehumanization of other persons, usually achieved through the redirection (and/or inversion) of the sex-drive. This alteration was aided and abetted (and also epitomized) by the bomb, which has long been associated with imagery that is both sexual and perversely pornographic. The perversity and pornography of the bomb might be said to inhere in its very structure and function: regardless of the political intentions of its inventors and handlers, the idea of the device itself flirts at the edges of moral obscenity. There are, unarguably, some inanimate objects in which the moral intents of their creators have been indelibly inscribed: the death camps, iron-maidens and thumb screws, booby-trapped children's toys. These constructs are not merely objects, nor have they been created merely to fulfill the objective of destruction; they reflect an almost psychopathic fixation upon the infliction of death and pain, either in ways, or upon a scale, that breaks beyond the boundaries of what we would designate as terrible. These objects imply a misanthropy so intense, and so vicious, that it can only be called obscene. While it would be wrong to claim that the bomb is a device designed to inflict such selective harm and sadomasochistically motivated pain, it has aroused, especially in the American subconscious, darkly erotic desires for final consummation in annihilation.

In discussing the bomb, Jacqueline Smetak strikes a decidedly Freudian chord when she posits that 'the aim of all life is death. Given that sex counters this by guaranteeing biological survival and that Eros, the life instinct, turns toward pleasure rather than destruction, any society intent on obliterating itself will repress or distort erotic tendencies. It will also repress awareness of whatever ... cannot be acknowledged, in this case, its desire to kill itself.'[13] While Smetak's conclusions may be somewhat extreme, the drives of both Eros and Thanatos can be rechanneled both to abet and reflect the technocentric super-state's trend toward systematic dehumanization of both its own warriors and their opponents. No other object has become a more effective medium for achieving these objectives than the atomic bomb, both in terms of the psychological force it exerts and in its significance as an icon that simultaneously embodies the yin and yang of both global destruction and sexual climax.

No narrative has made more use of this dualism in the imagery of the bomb than Stanley Kubrick's *Dr. Stangelove* (1964), significantly subtitled *Or, How I Learned to Stop Worrying and Love the Bomb.* Patrick Mannix, when considering the scene in which

General Buck Turgidson must leave his mistress in order to respond to the growing threat of global holocaust, comments on Turgidson's jocular exhortation to his sexual partner: 'Begin your countdown and Old Bucky'll be back before you can say "Blast off!" In making this statement, Turgidson is using the language of missile warfare to communicate a sexual message.'[14] Mannix is not alone in observing this connection between nuclear weapons and phallicism. Helen Caldicott's non-fiction tract *Missile Envy* (1985) contains this revealing passage:

> These hideous weapons ... may be a symptom of several male emotions: inadequate sexuality and a need to continually prove their virility plus a primitive fascination with killing. I recently watched a filmed launching of an MX missile. It rose slowly out of the ground, surrounded by smoke and flames and elongated into the air – it was a very sexual sight, and when armed with the ten warheads it will explode with the most almighty orgasm.[15]

It is naive to suggest that the ICBM, like the gun, is simply a phallic symbol, albeit writ in absurdly large dimensions. The more disturbing, and dehumanizing, aspect of atomic eros lies in the sheer scale of its destructive potential: the mere fantasy of using such a weapon is an obscene, and supremely ego-masturbatory, outrage against the ethics of most recorded civilizations and against human conceptions of scale and balance. If detonated, a nuclear bomb may vaporize millions, but even by merely existing, it shatters our ability to affix limits, to grasp the world in our accustomed framework of the finite. The fact of the bomb slays any functional faith in cognition, logic, law, order, morality, ethics. It is the ultimate symbol of transgression and violation, and to employ such a weapon could well be perceived as the ultimate and final act of self-gratification. Thus the bomb – being an absolute destroyer both of physical and conceptual boundaries and restraints – may propel us over the borders of acceptable, socialized behavior into bizarre mental topographies where Daliesque extrusions of megalomania, narcissism, and sadism burgeon chaotically, making an unmappable mockery of our own 'normative' experience. Burdick and Wheeler examine these obscene fascinations and perverse gratifications in *Fail-Safe*, when the nuclear expert Groteschele engages in a seductive death-dance with the sleek, predatory socialite, Eleanor Wolfe:

> When he described the Doomsday system, hinting that it was semiclassified, she closed her eyes for a moment and a light smile started at the corners of her mouth.
> 'Beautiful,' she said.
> Just that single word unaccompanied by an expression of horror or astonishment or dismay ... 'What makes you fascinating and what makes your subject fascinating is that it involves the death of so many people. Quite literally everyone on earth.' She paused a moment and then spoke savagely. 'Damn it, I wish I were a man and a man who could push the button. I would not push it, you understand that. But the knowledge that I could.' She shivered in her mink coat ...
> 'Why wouldn't you push it?' Groteschele asked softly ...
> 'Because I would die along with everyone else,' Evelyn Wolfe said.
> Her voice came to a queer faltering halt. Groteschele felt a very deep excitement. 'That is one statement you do not really believe,' he said with authority. 'Do you think that life is the most important thing to a person? ... Knowing you have to die, imagine how fantastic and magical it would be to have the power to take everyone else with you [...] all of them and their plans and hopes. And they are murderees: born to be murdered and don't know it. And the person with his finger on the button is the one who knows and who can do it.'
> The sound Evelyn Wolfe made was not a moan. It was the sound of wonderment that a child makes ... even if the child sees cruelty.
> 'Stop in one of those little side roads,' Evelyn Wolfe ordered.[16]

The predictable tryst that follows is, essentially, the Freudian death instinct and the countervailing sex instinct fused into one act of mutual consummation. The attraction and sexual encounter between these two characters is laced with metaphors not merely of predation, but of that most ravenous of all black magical eroticisms: vampirism.

This symbology of not merely *de*humanized, but *in*human, creatures is also reflected in another consistent narrative theme within

American nuclear war fiction: that of the linked onset of emotional detachment and madness. As the persons who must handle and control these weapons begin to adopt responses and habits that mimic the automata that are the true masters of such systems, their disassociation from intrinsic human needs, behaviors, and freedoms produces a fracturing of the self. Spencer Weart offers a particularly insightful review of the history of this thematic component in nuclear war narratives in *Nuclear Fear* (1988). Returning to the origins of American nuclear fiction in pre-Second-World-War science fiction magazines, Weart calls particular attention to Robert Heinlein's 'Blowups Happen' (*Astounding*, Sept. 1940), whose editor John Campbell described as being based on the very latest, laboratory-grade information (same issue, editorial). Heinlein's story concerns a nuclear power plant – candidly referred to as 'the bomb' by its attendant technicians – which is risky to operate, but has become an essential energy source. Weart points out that the core of Heinlein's story is not nuclear but psychological: 'Could any human being be trusted to operate such a plant, Heinlein asked, when one mistake might cause a catastrophe that would devastate thousands of square miles? Might not the pressure of the job drive an operator mad?'[17]

America's technocratic solution to this vulnerability was not merely the creation of machines that could do the work of men, but of men that could work like machines. The stoic, inflexible demeanor of totemic nuclear warlords such as General Curtis Le May (the architect of American nuclear strategy) and Admiral Rickover (the driving force behind the nuclear submarine) set a standard that not only 'trickled down' into their own branches of the armed services, but into the narrative characterizations of the de-libidinized machine-men who were charged with controlling and unleashing the arsenals of Armageddon. As Spencer Weart observes:

> Talk about steely logic, denial of feelings, and authorities controlling superhuman force could cast a chill shadow. Watching the SAC movies, a film buff with a good memory might have recalled scenes in *Frankenstein* or *The Invisible Ray* where the scientist ignored his woman as he relentlessly pursued secret powers. Russians wrote frankly of the American 'robot-soldier' who could 'drop atom bombs on civilian towns without shuddering.'

These inhuman standards of behavior and operational precision exacted the price that Heinlein anticipated in 'Blowups Happen.' Weart explains that

> Like the SAC airmen [...] nuclear submariners had to be impervious to sentiment if only because they literally held in their hands the keys to catastrophe. Magazine articles explained that the crews endured months locked in steel corridors, forbidden contact with their wives; 'after a while,' a reporter wrote, 'even their talk of sex stops.' The public did not know that Rickover's men, like Le May's, had morale problems and a high divorce rate. Nobody publicized the fact that in a typical year one out of every twenty-six missile submariners was referred to a psychiatrist and some had to be hospitalized for paranoid schizophrenia and other mental illnesses, a higher problem rate than in other branches of the Navy. The accepted image, rather, was of men who had made themselves into logical machines.[18]

The truth, of course, was closer to what Heinlein and others foresaw: 'Blowups' do happen, particularly in the highly reactive core of the human psyche. Indeed, the association between nuclear responsibility and diminished human affect and empathy began to become one of the most common tropes in nuclear war fiction. From the characters in the film version of *Fail-Safe* (1964) to the partially mechanical Dr. Strangelove, nuclear war narratives (particularly films) have brooded not only upon the physical mutation caused by the bomb, but upon the psychological degeneration of those who must live in proximity to it.[19] Whether or not such narrative representations are motivated by an authorial desire to investigate and/or critique such a phenomenon, they stand as mute testimony to the influence of the atomic bomb as a defining metaphor for the altered collective consciousness in post-World-War-II America.

Notes

1. *Threads*, a 1984 BBC-TV film directed by Mick Jackson, may well be the most devastatingly horrific of all the cinematic renderings of the nuclear nightmare. Its visual power and its effective synthesis of various rhetorical modes result in a singularly compelling – and arresting – anti-nuclear statement.

2. H. G. Wells, *The World Set Free* (London: Macmillan, 1914).
3. *The Day After* is a 1983, ABC-produced made-for-TV movie, directed by Nicholas Meyer.
4. Patrick Mannix, *The Rhetoric of Antinuclear Fiction: Persuasive Strategies in Novels and Films* (Lewisburg: Bucknell University Press, 1992), pp. 82–83, 137.
5. Ibid., p. 138
6. This essay is included in her landmark collection, *Against Interpretation* (New York: Dell, 1966).
7. Vivian Sobchack, 'The Violent Dance: A Personal Memoir of Death in the Movies,' *Journal of Popular Film*, 31 (Winter 1974): 2–14.
8. H. G. Wells, *The War of the Worlds*, in *The Time Machine and The War of the Worlds: A Critical Edition*, ed. Frank D. McConnell (1898; New York: Oxford University Press, 1977).
9. Whitley Strieber and James Kunetka's *War Day: And the Journey Onward* (1984) is a semi-documentary narrative. *Testament* is a film based on Carol Amen's 'The Last Testament' (1980).
10. The difficulty in assessing whether any single example of a 'post-nuclear frontier' story is essentially reasonable or ridiculous is usually problematized by the lack of specificity regarding the nature of the nuclear conflict which historically underlies the narrative scenario. This uncertainty rests in the imprecision of the term 'nuclear war' itself: this may refer to either a 'full strike' or 'limited exchange,' for both of these terms are imprecise. Is a single bomb an act of terrorism, or a limited strike? What total of inbound warheads or accumulation of megatonnage constitutes a 'full' strike? The assessment of whether the American 'frontier mentality' (and its implications of resilience, survival and rebirth) reflects extraordinary ingenuousness or reasonable optimism largely depends upon the scale of nuclear exchange posited by the author. Consequently, authorial imprecision or silence on this point not only allows, but invites, speculations upon the survivability of 'limited strikes.' This recurrent evasion has become one of the most subtle yet pernicious components of the 'popularization' of post-nuclear scenarios, a narrative-induced accommodation of inconceivable and unacceptable future histories.
11. Cf. Michael Crichton, *The Andromeda Strain* (1969).
12. I am neither claiming nor implying that it can be conclusively proven that Operation Olympic could have been made unnecessary through other methods. Many lengthy and compelling arguments have been adduced with suggest that the U.S. decision to employ the atomic bomb was not motivated by military concerns, so much as it was intended as a political statement to the post-war world and also as a chillingly utilitarian field test of an untried weapon. However, the Japanese propensity for suicide attacks did give military planners reasonable pause when considering any campaign to land upon and pacify the Japanese home islands, despite the utter ruination that had already been rained upon them. The strategic situation was further complicated by the Russian eagerness to enter the war with Japan at the eleventh hour, and to participate in both the conquest and

occupation of the home islands. Given the early signs that Soviet adherence to the Yalta and Potsdam agreements might be less than absolute, there was also considerable Anglo-American concern that, if given the chance, the Russians might attempt to forcibly convert Japan into a Soviet satellite state.

13. Jacqueline Smetak, 'Sex and Death in Nuclear Holocaust Literature of the 1950s,' in *The Nightmare Considered: Critical Essays on Nuclear War Literature,* ed. Nancy Anisfield (Bowling Green, Oh: Bowling Green State University Popular Press, 1991), p. 21.
14. Mannix, p. 154.
15. Helen Caldicott, *Missile Envy* (Bantam: New York, 1985), p. 319.
16. Eugene Burdick and Harvey Wheeler, *Fail-Safe* (New York: McGraw-Hill, 1962), pp. 123–4.
17. Spencer Weart, *Nuclear Fear: A History of Images* (Cambridge, Mass.: Harvard University Press, 1988), p. 82. Weart advances, and makes credible, the claim that '*Astounding*'s stories did more than any factual article to tell the meaning of fission'. There were other stories at that time which dealt with the same issue. One famous example is Lester Del Rey's 'Nerves', *Astounding*, 30 (September 1942).
18. Weart, pp. 151, 251–2.
19. Weart accurately observes that 'While the story had no robots running amok, it did have humans who would have been at home in a robot story' (*Nuclear Fear*, p. 276).

9

Pocket Apocalypse: American Survivalist Fictions from *Walden* to *The Incredible Shrinking Man*

GEORGE SLUSSER

Survivalism. To Americans today, the word suggests militias, armed preservation of personal freedoms in the wake of a violent (and much desired) collapse of the state. But there are much broader cultural resonances. Survivalism defines the traditional way America has dealt with its sense of an ending, conceived of the relation of the American individual to history and destiny. In what is commonly seen as its apocalyptic context, the central ideal of survivalism is simple. For when apocalypse is always now, thus the individual in perpetual training for the end, might this not be the way to defer this end perpetually? As with the title of the popular survivalist catalog *Loompanics*, panic ever looms. In this catalog, the citizen finds the self-help manuals that enable him to prepare. These are the means whereby each can, if only for himself alone, foreshorten secular and divine history, whereby each stretches individual life into an endless 'now' of preparation, in a forever war against Armageddon.

However strange this idea, it remains in American culture a central response to growing forces of technological and social change from Emerson down to modern SF. One cannot deny the real and growing presence of holocaust and apocalyptic menace in this century. Nor should one ignore the growing presence of so-called multicultural activism in America, for whom the individualist imperatives of the survivalist are anathema. Yet I would argue that the very form of this challenge, in its utopian and millenarist nature, is itself determined by a same cultural fixation on apocalypse. In professed counter-apocalyptic novels like Ursula Le Guin's *Always*

Coming Home (1985), there is also a like desire to stretch now into eternity, if only in this case to turn the biblical 1000-year reprieve into little more than another million-year picnic. Informing the landscape of American culture from the nineteenth century to today, I see a close relationship between apocalyptic and millenialist thinking. Indeed, these impulsions, issuing from a common source in Thoreau's *Walden*, offer opposite-seeming and yet quite similar responses to ever-gathering forces of material change. The mirror that reflects their deeper likeness is Thoreau's peculiar form of survivalism. For in *Walden*'s survivalist experiment Thoreau, rewriting in his imagination nature's history in Biblical terms, creates a refuge where the American individual, in order to abide, must perpetually act the role of stranger in his own land. Thoreau is the first to explore what it means, in the survivalist sense, to 'imagine' disaster. For it is by means of such acts of terminal imagination, such end games, that we learn how to put apocalypse in our pocket, and in corollary fashion, how to send away for mail-order millenia. I focus here primarily on the pocket apocalypse. To suggest the range and scope of this gambit, I will look briefly at three narratives: Thoreau's 'peaceful experiment'; Hemingway's response to the war to end all wars, 'Great Two-Hearted River'; and Jack Arnold and Richard Matheson's imploding world in *The Incredible Shrinking Man*.

WALDEN AS SURVIVALIST EXPERIMENT

The intellectual context of *Walden* is Emersonian 'self-reliance'. In this curious vision, we see society 'everywhere... in conspiracy against the manhood of every one of its members'. The individual is less openly than passively rebellion against social constraints: 'It is easy to live in the world after the world's opinion; it is easy in solitude after our own. But the great man is he who in the midst of the crowd keeps with perfect sweetness the independence of solitude.'[1] Such a man, free and entire of himself, is not like Donne's island, in need of being brought into contact with the main. Instead he becomes, in the covert act of cutting social ties, a center, not just of his world, but of *the* world, as a still place, a 'now,' in the midst of temporal change: 'A true man belongs to no other time and place, but is the center of all things.' This declaration of self-centeredness is at the same time self-empowerment: 'No law can be sacred to me but that of *my* nature.' But for this law to abide, the self that proclaims it has to continue to exist, even beyond the existence of that nature of which he is presumably a part: in other words, to *survive*. As Emerson puts it, 'Power ceases in the instant of repose; it resides in the moment of transition from a past to a new state, in the shooting of the gulf.'[2]

Walden is, to use Thoreau's words, an 'experiment' in survival*ism*, that is, an intellectually contrived situation, voluntarily begun and ended, a training exercise in Armageddon. Thoreau here offers several crucial imperatives, which must serve as survivalist blueprint for others to apply. The first reveals that, though the impulse to survive may come more from imagination than from the pressures of history, the thought experiment must be lived out in a concrete, physical manner: 'If you have built castles in the air, your work need not be lost ... Now put foundations under them.'[3] The second, under the aegis of making this act of willed, non-necessitated presonal survival a material act, is a response to the Cartesian *cogito*: 'Let us spend one day as deliberately as Nature' (1493). Here, instead of proposing to separate self from world (we remember Emerson's man alone *in the world*), Thoreau bids the individual to conflate his material self (as a mind now firmly *located* in the center of the individual body) with various space-time extensions of history and deity. Indeed, for Thoreau, the body bears not only a symbolic relation to the deity ('every man is the builder of a temple, called his body, to the god he worships' [1574]), but can be literally an 'epitome or abstract' of that God. Paradoxically, it is by tending to the here and now of one's physical present that one reaches out to the end of eternity; in doing so one draws apocalypse, in undulatory fashion, back to the dynamic center of the self: 'In eternity there is indeed something true and sublime. But all these times and places and occasions are now and here. God himself culminates in the present moment' (1493).

The survivalist experiment of Walden Pond involves not only establishing this undulation between the apocalyptic circumference and 'hereness' of self, but controlling it through fictions which themselves are materializations of what Thoreau calls acts of 'imagination'. The centripetal impulse is given by the call to 'simplify'. This, in one sense, operates as a materialized form of Cartesian doubt, a conscious and deliberate divesting the self of all superfluous social and sociohistorical ties – an act of 'voluntary poverty'. Thoreau first enacts this on the etymological level, by taking concepts like 'poverty' back to their concrete and literal origins – 'small means' in this case. An example is the cluster of abstract words associated with 'business' – 'goods', 'commerce', 'economy'. By saying he is going to Walden to 'transact private business', Thoreau divests the word of its socio-economic connotations; for 'busy-ness' here will be purely personal activity. And if any 'goods' come from this activity, they are 'right and proper' only for him who made them. For Thoreau, the very act of describing his survivalist fiction of Walden Pond in economic terms becomes a powerful means of loosening the hold of socioeconomic ties on his private existence: 'I too had woven a kind of basket of a delicate texture, but I had not made it worth any one's while to buy them ... and instead of studying how to make it worth men's while to buy my baskets, I studied rather how to avoid the necessity of selling them' (1441). The only meaningful 'commerce' one can have ('*com-mercari* > trade together > trade – tread') is with things one uses in one's own handcrafted world. Such is the physical passing of the axe Thoreau uses to clear the land at Walden: 'The owner of the axe, as he released his hold on it, said that it was the apple of his eye; but I returned it sharper than I received it' (1456). Commerce (here 'selling') has, in being simplified, recovered its fundamental sense of physically taking hold, or loosening grip on, a physical object, one whose sole 'value' lies in being sharpened by use.

The centrifugal impetus, moving out from Walden into futurity, is given by the 'Fitchburg Railroad' that 'touches the pond about a hundred rods south of where I dwell' (1505). The railroad, as construct of a new technological society, both fascinates and terrifies Thoreau. Its regularity threatens to supplant that of nature (farmers set their clocks by its whistle); its power is a juggernaut, carrying all mankind away from its center in self-reliance to some fateful terminus: 'We have constructed a fate, an *Atropos*, that never turns aside' (1507). The only way, however, that the narrator can ride this railroad to the end of time, rather than be ridden by it, is to make it the vehicle of his imagination. In doing so, he sets in motion the undulation between center and a newly discovered circumference. Turning on the presence of the railroad, the call to simplify ('if we stay home and mind our business, who will want railroads?') becomes simultaneously a call to amplify: 'if railroads are not built, how will we get to heaven in season?' The 'tool' the survivalist uses to secure his freehold in the face of this surging fatality is not technology itself (despite Thoreau's many descriptions of building a dwelling and making clothes, a garden and such) but the *imagination* or fiction of technology that his railroad, figurative and literal at one and the same time, comes to be. It is only by imagining we can ever get to heaven 'in season', whether the train be moving along the path of *chronos* or that of *kairos*, that we come to realize we are already in heaven, now and forever in the survivalist present.

'Time is but the stream I go a-fishing in.' Thus the narrator seeks to condense the space-time continuum into a single act in the survivalist 'now.' But as the railroad's presence demonstrates – an iron 'stream' that cuts through his world and cannot be fished in – survival means further accommodation, and this is achieved only by setting up, at a point where technology intersects with nature, an undulating rhythm between a new center and circumference: the end of collective time, and the timeless present of the personal consciousness. The narrator forges such a nexus when, in describing the thaw in the bank of the cut made by the railroad, he literally conflates the physical power of dynamite with the force of his own intellect as it recreates the interaction of organic and inorganic forms in turn caused by this intersection of technology and nature. For Emerson in 'Nature' (1836), the act of creating such an undulatory center remains an ethereal one: 'I become a transparent eyeball: I am nothing, I see all.' For Thoreau, however, 'the intellect is a cleaver; it discerns and rifts its way into the secret of things' (1494). Mind becomes the immediate physical analogue of the railroad cut. It allows its own reflections to blend with the riot of thawing forms, and thus bridges the gulf between living and inert matter: 'What is a man but a mass of thawing clay? The ball of a human finger is but a drop congealed' (1628). These are neither figure or metaphor, but analogies that literally place the individual, mind and body, at the center of a process that, in this description, extends outward to encounter cosmos itself: 'Who knows what the human body would expand and flow out to under a more genial heaven?' (1628).

The answer to this question, in the description, is an expansion that encompasses first apocalypse, then millenium, as terminal forms of inertia that themselves must be rendered vital again. In the turmoil of the thaw we witness the enfolding, within the compass of a single instance of human perception ('it seemed that this *one* hillside illustrated the principle of all the operations of nature [p. 1628]'), of a full-fledged apocalypse. We experience, more accurately, a concretization (or immediate rendering in the 'now') of the predictive abstractions of Biblical apocalypse. In the place of 'cataclysm', there is *this* literal washing away or flood from the thaw. Rather that 'catastrophe' (an 'upheaval'), we hear that 'there is no end to the heaps of liver, light, and bowels, as if the globe were turned wrong side out' (p. 1628). On one hand, the throes of this inscribed pocket of space-time seem to expand in an instant to universal proportions. Encompassing all the earth and natural history, they 'will heave our exuviae from their graves'. There is a clear terminal moment, where not only 'the forms which this molten earth flows out into,' but 'the institutions upon it are plastic like clay in the hands of the potter' (1629). And yet, on the other hand, this remains at all times a private revelation, the apocalyptic moment circumscribed as personal experience. The Biblical portents themselves are made to become, through the process of relating this experience, signs of individual duration in an inexhaustible plenitude of nature: 'I love to see that Nature is so rife with life that myriads can be afforded to be sacrificed and suffered to prey on one another [...] and that sometimes it has rained flesh and blood!' (1635)

Once Thoreau creates his pocket apocalypse, he moves on in the final pages to address the millenialist temptation. Paradoxically, this is more threatening yet to the survivalist, because it offers, as part of the terminal bargain, a thousand years of utopian calm, where the self-reliant individual must endure an Arcadian world of peace and plenty. This world, it seems, is collective and global. Whereas Emerson tells us that the soul is no traveller, for the millenialist all are now free to circulate through any clime or species. Thoreau makes it clear, however, that voyaging through space and time in search of this brave new world is 'only great-circle sailing'. The millenium he proposes instead, from survivalist Walden Pond, is a mail-order one; staying put, we need only send for brochures upon which to construct our personal millenia. These occur, not at the end of time, but in some inexhaustible 'now' at the center of the natural world.

Walden ends with a wish that resonates today with the American survivalist:

> Who knows what beautiful and winged life, whose egg has been buried for ages under many concentric layers of woodenness in the dead dry life of society [...] may unexpectedly come forth from amidst society's most trival and handselled furniture, to enjoy its perfect summer life at last!

The fulcrum here – the place of private metamorphosis upon which the relation between collective destiny and personal freedom turns – is the word 'handselled'. One meaning is 'trivial', signifying the end product of the dead, dry life of millenialist utopianism. Another however, is the 'inaugural token', a thing signifying good wishes offered at the beginning of a new phase or state. In this context, to inaugurate the 'golden age' is, again, little more than the single individual's ability to lay his egg, hoping for survival and even transformation, in the warp and woof of social progress and destiny.

SURVIVALISM AND THE IMAGINATION OF DISASTER: NICK ADAMS TO THE INCREDIBLE SHRINKING MAN

If *Walden* looks forward to the persistence of survivalist fictions in American culture, the twentieth century offers its own apocalyptic challenge to the idea itself of survival. There have been two decimating wars, the Nazis, and finally the threat of nuclear annihilation of the human race and earth itself. Thoreau's enclave of survival seems on the verge of being submerged by real disaster. If anything American culture, from the war fictions of Hemingway to the science fiction disaster films of the 1950s, seems to revel in mass destruction. Yet there are notable exceptions amidst these totalizing visions, works in which the survivalist response subtly reasserts itself. Let us look first at Hemingway's short story 'Big Two-Hearted River'.

Though Hemingway is on record as finding *Walden* unreadable, 'Big Two-Hearted River' echoes Thoreau's remark that 'Time is but the stream I go a-fishing in'. In the narrative chronology of the Nick Adams stories, this tale recounts the protagonist's retreat from the juggernaut of World War I to a place where he is free to re-imagine disaster, but now in a setting he can control, and whose center is formed by his personal skill at surviving.[4] Indeed, the fact that Thoreau's 'experiment' at Walden Pond is a subliminal presence in Hemingway's story may account for the tendency on the part of critics to read its literal, functional prose as symbolic, especially in terms of Biblical symbolism depicting Edenic beginnings and apocalyptic endings.[5] On one hand, the narrative recounts the living, in Thoreau's sense of 'deliberate', of two single days, recounted one by one, as Part I and Part II. But the telling of simple things summons at every instance radical flights of the critics' imagination to the ends of time and space. There is, as with Thoreau, an undulating nature to Hemingway's narrative, with the careful recounting of each physical detail, the center of personal experience creating resonances on its symbolic circumference. Furthermore, from this center the imagination, on day one, is drawn inward toward the beginning of things, suggested in the narrative by hints of the millenial language of the Golden Age. On day two, it is drawn in the opposite direction, outward to the apocalytic end of time. It is the undulation of these 'imaginations', as if they were intersecting cones, that sustains Nick's personal survival.

The narrative begins with the train moving out of sight and Nick sitting down in a place described in bluntly literal fashion, but which simultaneously resonates with apocalyptic symbolism: 'There was no town, nothing but the rails and the burned-over country.' If Thoreau's imagination raced ahead of his rails ('If the railroad reached around the world, I think I should keep ahead of you', *Walden*, p. 1464), Nick has already reached the terminus. A second spatio-temporal vector, however, moves through the burned-out land – the river itself: 'The river was there [...] Nick looked down into the clear, brown water, colored from the pebbly bottom, and watched the trout keeping themselves steady in the current with wavering fins.' (p. 209) As we shall see during the day Nick spends fishing this river, the place of experience is at one and the same time a personal and figurative place. The river is 'two-hearted' perhaps (the story's title is never explained) because it offers Nick two directions in which to move, and the reader two ways in which to exercise his or her apocalyptic imagination. The first is inward, a millennial movement downward through the land of soot – Nick looking into water that is both 'brown' and 'clear'. The second (horizontal to the first day's vertical movement) follows the river's flow outward, toward the perilous fishing of the swamp, a place where apocalyptic figurations are concretized as Nick's sense of a personal fate. At this moment, human 'survival' is one with Nick's retreat to fish another day.

Thoreau defines living the day in the apocalyptic current as either despair of life and pursuit of 'a descending and darkening way', or the belief that 'each day contains an earlier, more sacred and auroral hour than he has yet profaned' (p. 1488). On his first morning, Nick pursues this earlier more sacred hour within the profaned day. He picks up one of the soot-covered grasshoppers ('locusts' of the apocalypse) and turns it over to see if its underside is also black, then tosses it in the air: 'Go on [...] fly away somewhere.' He later notices, on the sunset side of this same day, that these same flying insects draw feeding trout out of the water: 'As far down the long stretch he could see, the trout were rising, making circles all down the surface of the water' (p. 214). The rhythms of nature involve species feeding on one another, but Nick discovers in this spectacle, if only provisionally in his march across the day, something millennial. As he moves from the ashes toward the river, 'underfoot the ground was good walking.' At the meadow on the river's edge he sets up camp: 'It was a good place to camp, he was here *in the good place*' (p. 215). In the undulatory rhythm of these simple sentences, the word 'good', connoting a moral absolute, is also never more than what we make of the place we are: 'There were plenty of good places to camp on the river. But this was good.' The tent he sets up, again undulating between figurative and literal, is a place of light in darkness: 'It was quite dark outside. It was lighter in the tent' (p. 215). The space Nick creates is, simultaneously, Arcadian and ephemeral.

The next day, the fishing expedition, sees Nick moving away from temporary Arcadian harmony with nature toward an encounter with apocalyptic terror. That morning he finds 'plenty of good grasshoppers' by turning over a log and discovering what the narrator (very close to Nick's thoughts) calls 'a grasshopper lodging house'. Again, working himself into the natural rhythm of things, Nick has a millennial moment of harmony within a natural world otherwise violent and brutal. Though the hooking of the 'fifty medium browns' he captures is cruel, he avoids greater senseless cruelty by not having to crush a much greater number by flailing around with his hat to catch them in the open.

In his first strike, while fishing in shallow water, Nick retains this 'ecological' relation with natural forces. Taking a small fish, he unhooks it and puts it back in the stream. During this operation, 'he had wet his hand before he touched the trout, so he would not disturb the delicate mucus that covered him' (p. 225). But now, lured by the 'big trout', he moves into 'smooth and dark' water between meadow and swamp. What he catches is a Leviathan, that breaks his line and gets away: 'He had been solidly hooked. Solid as a rock. He felt like a rock, too, before he started off. By God ... he was the biggest one I ever heard of' (p. 227). This encounter is a rupture, leaving behind an 'angry' fish with a hook in his mouth. Nick must leave the stream or be overcome: 'Nick's hand was shaky. The thrill had been too much. He felt, vaguely, a little sick, as though it would be better to sit down' (p. 226). Pulling himself together, he returns to the stream and, now as master fisherman, takes two big trout in succession. The while, however, he has been moving toward the swamp. Again he must stop short: 'In the fast deep water, in the half light, the fishing would be tragic. In the swamp fishing was a tragic adventure' (p. 231).

Nick Adams is not a talker like Thoreau's narrator at Walden; he is a doer. His words, in direct speech, are few, inarticulate, and generally banal. Nor does the third-person narrator ever rise above the level of this plain speech. Neither have the wide range of cultural memory that Thoreau displays; indeed, Nick's personal memories do not reach beyond another, earlier fishing trip on the Black River (Thoreau's narrator would make the connection, left to the reader here, with Acheron). Given the fact that the best adjective Nick can muster for his millenial moments in nature is 'good', the word 'tragic' that appears here takes on uncanny power. As a word, it simultaneously carries the force of all mankind's apocalyptic condition, and confines it to the easily controllable situation of a day fishing in the river of time. As with the word 'good' the first day, here the 'tragic' is evoked only to be immediately domesticated to Nick's personal situation. If Nick now appears a God in his universe, killing trout only to seem to resurrect them ('He washed the trout in the stream. When he held them back up in the water they looked like live fish'), he is still only fishing in this stream, at this hour and day. And yet, if Fate rules out there, at the tragic end of things, Nick at his center is free here to choose his moment: 'There were plenty of days coming when he could fish the swamp' (p. 232).

Science-fiction films of the 1950s develop under the menace of total annihilation of the human race. With the threat of atomic annihilation looming ever larger, these films generate increasingly radical visions of collective apocalypse, apparently unredeemed by any possibility of personal survival. Yet, at the heart of this cinematic holocaust, one film, Jack Arnold's *The Incredible Shrinking Man* (1957), strikes a curious counterbalance. Sandwiched between *Invasion of the Body Snatchers* (1956), which narrates the triumph of the Beast of Cold War Communism, and *On the Beach* (1959), where the audience gets what it has been waiting for across this decade – total nuclear holocaust – Arnold's 'survivalist epic' offers a strange, disaster-driven variation on Thoreau's *Walden*.

The scope of *The Incredible Shrinking Man* is, in the manner of *Walden*, as narrow as the everyday existence of a common man and his 'nuclear' family in his suburban house, and as broad as the end of space and time as we know it. What is more, this end is clearly, as the credit sequence reveals, the product of atomic holocaust. As the credits appear on a bare, greyish screen, accompanied by a Ray Anthony trumpet solo, on the left side a white human silhouette two-thirds the height of the screen begins to shrink. On the left side, a small white mushroom cloud simultaneously starts to grow until it dwarfs the shrunken human figure. An inverse proportionality is set in motion that cannot stop until the human form reaches the zero point of annihilation. This, however, is but a premonition, for before the figure visually disappears, the credits end.

The opening shot is a blank seascape, empty of any human presence except the voice-over of the narrator. He offers to tell the incredible story of Robert Scott Carey, and promises to do so from an authoritative point of view: 'I know the story, because *I am Robert Scott Carey*.' This story is 'incredible' because it is simultaneously the story of a man who has shrunk to nothingness, and a testimony (because the narration is in the past tense) of this man's impossible survival. From the outset, looming absence, the inexorable annihilation of self, is juxtaposed with continuous presence of that same self. This paradox can be accepted to the degree that it is subsumed by an undulatory rhythm such as that which is posited of Emerson's 'transparent eyeball': 'I am nothing; I see all.'

The story of Scott Carey is narrated, both in terms of visuals and spoken narrative, in an unbending straight line. The effect, on this level, is one of a man being carried by Thoreau's engine, 'an *Atropos* that never turns aside', to the zero point of total destruction. The significance of this is clear when we compare the film with Richard Matheson's novel from which the screenplay is adapted. In the novel, Matheson narrates his story with a series of flashbacks and flash-forwards which, if intended to complicate things on the level of *récit*, confound the logic of the shrinking man postulate. In contrast, the film's linear telling of Scott Carey's journey creates a near-epic sense of inevitability.[6] In the opening scene, the camera pans up from its myopic focus on the empty sea to the deck of a boat occupied by a sunbathing Scott Carey and his wife. Suddenly, sweeping over this scene of domestic harmony, as his wife goes below to fetch a beer, is Nemesis, in the form of a mysterious white mist that covers him with flakes, then passes, unseen by anyone else. Scott then returns to normal daily life in his suburban home, where everything is in 'proper' order and place. Gradually however, by increments, signs appear that this equilibrium is shifting: his wedding ring falls from his finger, his clothes become too large. The doctors who examine him are first incredulous: 'human beings don't shrink'. Measurements, however, reveal he is diminishing in height and width. It is concluded that irreversible transformation of cellular structure has occurred. Yet were this so, Scott's form itself would change. As it is, what changes is simply a matter of scale. Inside a human form always intact yet always diminishing, Scott undergoes progressive alienation from what we call the human condition.

From now on, this dual rhythm – on one hand the terror of unending shrinking, on the other a periodic 'refocusing' of Scott Carey's form in relation to the viewer – becomes a function of the visual experience of the film itself: its 'special effects'. These exist to make us see Scott becoming tinier – first in relation to his shirts, then to chairs and furniture in his house, to his wife and other human beings, and finally to the very writing implements (pencil and paper) with which he struggles to record his journey toward the vanishing point. These manipulations of perspective, however, call attention not only to the relative nature of things as perceived, but (even more strongly) to the psychological needs of the viewers who not only perceive Scott's but must relate to Scott's adventure. If Scott really shrinks visually in relation to objects around him that remain in the viewer's frame of reference, he becomes a pinprick on the screen, and his adventures unviewable, hence irrelevant to us in our 'normative' world. If intellectually we know he is getting smaller, visually we ascertain the opposite – that his world is getting bigger, and Scott, at the center of this expanding world, remains the same – 'our' size.

The scene of Scott in the huge chair is pivotal. Up to this point, we have seen him, though shrinking, in more or less 'normal' relation, proportion-wise, to other human beings on the screen. Now the camera observes a conversation between Scott and his wife from behind the chair in which he is sitting but (because of the camera angle chosen) would not normally be seen. We then cut to a frontal shot and discover, from a high angle as if looking down from the wife's now-towering position, a much-diminished Scott sitting in what we thought a normal chair. The cut now is to Scott's point of view, placing the camera at an extreme low angle, from which the wife is viewed. Now, after this oscillation of perspectives, the camera focuses again on the protagonist, reframing him in relation to the viewer as a 'normal' human figure, but now sitting in a giant chair. The viewer has seen Scott get smaller; at the same time, however, that same viewer realizes that the photography that gave him this visual experience is 'trick' photography, and as such a deviation from the norm. The same is true for the props – the chair that first looked to be a thing in its 'proper' place and hence a measure of Scott's diminution, is revealed, once the narrative focus returns to Scott, to be an oversized set, an artificial construct 'bigger than life'.

Scott's doctors give him hope, that his condition has stabilized. Alienated, however, by his size from wife and home, he goes out in the night and meets a midget girl. Because the girl is played by the same actress as his wife, the viewer's intellect says one thing (this is a relationship played out on a different scale of being, freakish but still human), while the eyes say another. For once Scott and his friend are seated in the giant bar furniture, the camera reframes them in relation to the viewer – as two 'normal' people talking to each other – and the rest recognizes as oversize sets.

Given these undulations of perspective, the viewer constantly asks: Is Scott shrinking in relation to the world, or is the world growing bigger around him? For the process does not stop. The idyll with the midget girl may lull us momentarily into *thinking* that to shrink is to find human worlds within worlds – worlds we can reframe intellectually as well as visually. The camera, however, offers no such Pascalian mean. As Scott continues to shrink, it must stage more visual surprises, more radical disparities between a small human form and 'familiar' objects turned monstrously large. And it must, in order to tell the story visually, resort to increasingly 'incredible' reframings. The real drama of this film becomes the viewer's apprehension – how will the camera reframe (because we know Scott still exists) the time and place where he must vanish?

As shrinking continues, Scott sheds all but a few vestiges of his once-familiar world. Ordinary things are no longer 'other', they are menacingly so. The visual undulations between center and circumference become increasingly vertiginous. We see Scott sitting alone among 'normal' furniture, the medium shot framing him in proportion to things around him and to the viewer. Then we cut to a long shot, which reveals Scott's new living room to be part of a doll's house. From our familiar sense of doll houses, we judge Scott to be the size of a mouse. The cat he could pet earlier in the film, of necessity, is now his predatory enemy. But again, the camera alternates between blow-ups of the cat, and size-diminishing long shots of Scott. Trick photography is able to merge these two perspectives in single frames, and yet again the trickery only calls attention to the camera's abiding need to reframe the shrinking man.

Escaping from the cat, Scott falls down the basement stairs to the floor below. Because his wife thinks the cat has eaten him, his rupture with the human world is now total and irreversible. The oversized sets and extreme low angle shots turn this common staircase into an unscalable Everest. For Scott, the basement floor becomes *the* primal terrain of survival, on which his first and last man confronts an entirety of nature. Intellectually, 'shrinkage' becomes devolution, mankind progressively stripped of civilized attributes, forced into atavastic combat for food and shelter with monstrous creatures that turn out to be common insects and spiders, which because of his ever-diminishing size he is destined to lose.

As the camera reframes Scott for a last time, however, the struggle it visualizes becomes an evolutionary one – the human figure we watch is, in our scale of things, 'normal', and the blown-up image of the spider, by comparison, every bit a monster, and its defeat a supreme act of human courage, ruse against force.

The ultimate undulation of the film however is not devolution and evolution, but apocalypse and individual 'survivalism'. If in *Them!* nuclear radiation caused insects to grow, here it causes a human to shrink. The result, however, seems the same – small mankind fighting primal creatures resurrected anachronistically in the atomic age, near the end of civilized time. The nuclear apocalypse is, effectively, a collapse of Alpha and Omega, the conflation of beginning and end. And yet, here in this cosmos on a basement floor, endings and beginnings alike, as circumference of human history, continue to undulate with the central individuality of Scott Carey. As future and past converge in Scott's now, the survivalist paradox of this film becomes clear: the only way Scott can preserve his existence is by losing it, in passing through the needle's eye of nothingness. To reframe Scott at the vanishing point – where the end rejoins the vacant seascape of the opening shot – is again to place a frame around the void. In visual terms, however, this end is not exactly the same as the beginning. There voice and natural world are separate, the camera must track from the empty sea to Scott's body on the boat. Now as his form recedes *within* a frame of the natural world, the fact that he continues to speak tells us he has passed beyond our (the viewer's) range of vision. At this point, however, instead of a fixed shot on nature, the camera begins to move in relation to the frame in which Scott has vanished in a way that suggests a transfer of his presence to a new mode of existence. Things turn inside out, and what held the center of our attention as human form that existed by being seen, now frees itself to become that which sees instead.

In the final scenes, visual reframing, which thus far had allowed Scott to survive at the normative center, yields to what is an act of Thoreau's apocalyptic imagination, where both shrinking and survival become premises, abstractions that operate no longer on the scale of physical 'reality' but of that of a power to imagine endings and beginnings. Such reframing is the result of Scott's ability to redefine himself in terms of survival. The moment of redefinition comes as, about to give combat to the spider, he states, 'My brain was still a man's brain.' If he has actually shrunk in relation to the viewer's physical world, this statement is absurd. For a mouse, let alone a flea or mite, does not have the neuronal mass of the human brain, hence could not devise the strategies needed to survive that Scott clearly does. But now, in a rhythm that reverses the relation of human form and mushroom cloud of the title sequence, Scott's physical being can only be said to shrink in inverse proportion to the expansion of his (increasingly 'imaginary' or circumferential) sense of self survival. He uses human ruse to kill the spider, and then begins an ascent, up the dirt slope formerly guarded by the creature, and out a hole in the mesh screen he was earlier too large to pass through, onto the grass under the night sky. Before, the result of Scott's shrinking was entrapment (the basement stairs he could not climb, his insect-like size which made him inaudible to wife and brother, whose monstrous shoes inadvertently almost crushed him). But now, further shrinking becomes a means of liberation, as he slips through the mesh into the wide world amid the wonders of grass, dew, 'the flaring pin points of the stars'.

At this moment of ending, a significant difference between film and novel helps us see the implacable survivalist logic of the former. If by its logic the film must end in Emerson's paradoxical situation – where being nothing and seeing everything, center and circumference, coincide – Matheson's novel swerves away at the last minute toward a millenialist reprieve. At the presumed vanishing point, Matheson's Scott Carey encounters one of Pascal's 'two infinities' – the infinitely small:

> Why had he never thought of it: the microscopic and submicroscopic worlds? That they existed he had always known. Yet

> never had he made the obvious connection. He'd always thought in terms of man's own world [...] For the inch was man's concept, not nature's. To a man, zero inches mean nothing [...] But to nature there was no zero. Existence went on in endless cycles [...] He would never disappear, because there was no point of nonexistence in the universe.[7]

If there is the possibility of individual survival in this new world, this Scott seeks more. His hope, it seems, is to rebuild the very world Thoreau sought to simplify, to discover intelligent life, then human companionship, and eventually human society: 'There was food to be found, water, clothing, shelter. And most important, life. Who knew?'

Pascalian infinity is made the place of utopian-millenialist reprieve here, and the ending of the novel differs only in number and size from that brave new world born at the heart of apocalypse in a film like *When Worlds Collide*. Arnold's film patently refuses this. In the film, Scott's final words are: 'To God there is no Zero. I exist.' Yet, in contrast to such a quasi-Cartesian revelation, the final images of the film offer a rigorously Emersonian dynamic. Throughout Scott's shrinking, the camera has sustained a subtle undulating rhythm between disappearance and survival of the human center by manipulating the relation between Scott and his various circumferential 'frames'. Now that same camera, affirming itself as the location of Scott's continued existence, literally reenacts Emerson's dictum. From an extreme high angle, it travels up and away from Scott as he stands amid the drops of dew and blades of grass, until he is indistinguishable from them, invisible to the camera eye. At this juncture, the camera rotates 180 degrees and, pointing at the 'pinpoint' stars, begins to travel toward them, creating a new center that, this time, must become nothing so as to see all, to pass unimpeded to the end of the universe. Here, in equally emblematic form, is a survivalist response to the dynamic of the opening credits. There, the inverse relation between human form and mushroom cloud, once set in motion, appeared to be inexorable. Now, at the point where Scott is no longer seen, what appears is a fulcrum, a place on which the relation between power and form turns that, for writers from Emerson and Thoreau down to Hemingway and science fiction 'libertarians', defines the nature of individual survival.

Survivalism, then, is not an anomaly of modern libertarian thinking in the United States. On the contrary, it is a deep-seated cultural response to the persistent apocalyptic imagination that appears to have haunted even the peaceful 'experiment' of Walden Pond, in a nation where even the most orderly withdrawal from social existence is marked by quiet desperation, ordinary life at every instant on the edge of the abyss. Thoreau's tract offers a training manual for the urban nightmares and armageddons of the next century. It is easy to trace this current in the real-life fantasies of *Loompanics*, or in the SF extrapolation of a work like Robert A. Heinlein's *Farnham's Freehold*. Less obvious in Hemingway, or in a low-key film like *The Incredible Shrinking Man*, Thoreau's survivalist paradigm offers a response to the imagination of future wars, as it does to the millenialist temptations, the 'ecological' halt, that we find in Nick Adams's world and that of Thoreau alike. Survivalism, it seems, is a prime imperative of American culture, one that drives fictions on all levels. Far from being the purview of 'fringe groups', or of a 'fringe' literature like SF, survivalism, with sacrosanct roots in *Walden*, shows us that America has been a science fiction world from its beginning, a world whose manifest destiny was to create this literary form that has come to dominate its cultural landscape as it faces the new millenium.

Notes

1. Ralph Waldo Emerson, 'Self Reliance', in *Anthology of American Literature*, I, edited George McMichael (New York: Macmillan Publishing Co., Inc.), 1974, p. 1322.
2. Ibid., p. 1329.
3. Henry David Thoreau, *Walden*, in *Anthology of American Literature*, I, ed. George McMichael (New York: Macmillan Publishing Co., Inc.), p. 1639.
4. *The Short Stories of Ernest Hemingway* (New York: Charles Scribner's & Sons), 1938, pp. 209–32. All citations are to this edition. The story was originally published by Hemingway as the last story in *In Our Time*. In his memoirs *A Moveable Feast*, Hemingway says that 'the story was about coming back from the war but there was no mention of the war in it.' (*Feast*, Scribner's: New York, 1964, p. 76).
5. An example is found in Joseph M. Flora, *Hemingway's Nick Adams* (Baton Rouge: LSU Press), 1982: 'By dividing his story into two parts, and so labelling it, Hemingway emphasizes the two-heartedness of the river, and he also reinforces the rhythm of Genesis. The story will give us one day exactly. Nick is like Adam at creation, for sleep has – the structure indicates – brought no disturbing dreams.' (p. 164)

6. Richard Matheson, *The Shrinking Man* (New York: Fawcett), 1956. The book opens with a short chapter that presents the 'first cause' – the boat and the encounter with the fatal 'mist', and the film's opening scene. The second chapter cuts abruptly to Scott on the basement floor, in the midst of mortal combat with the spider, which is the film's penultimate scene. The narrative will now cross-cut back and forth between this alpha and near-omega. The purpose of this formal device is to present what seems to be the iron limits of one adventure, to let the journey of the shrinking man from here to there play itself out, only to reveal to the reader in the end that there is more, a world where Scott steps beyond his apparent fatality to new (and marvelous) possibility. In the film however, what is significant is that fate and survival are both present, simultaneously, from beginning to end, interrelated in undulatory manner in like manner to Emerson's elliptical pair power and form ('human life is made up of two elements, power and form'), with fate as the circumferential element or 'form', and survival the power that continuously fills and resists its formal container.
7. Matheson, p. 188.

10

'An Unrehearsed Theatre of Technology': Oedipalization and Vision in Ballard's *Crash*

NICK DAVIS

> The technological is never grasped except by (auto) accident.
>
> Baudrillard[1]

> It is clear that Freud's classic distinction between the manifest and latent content of the inner world of the psyche now has to be applied to the outer world of reality. A dominant element in this reality is technology and its instrument, the machine.
>
> *The Atrocity Exhibition*[2]

Ballard's novels of the early 1970s offer themselves as the revelatory annals of a world which has turned apocalyptic, and where received, normative systems of representation operate as a blocking of this fearful knowledge. *The Atrocity Exhibition* (1970) and *Crash* (1973) explore shared conditions of life which are catastrophic in ways that characteristically elude consciousness. What is restrictive in the ordinary workings of consciousness becomes clear through the novels' direct dealings with scenes and situations of catastrophe, whose function here is to disrupt the normalizing vision – a field made up of what Ballard elsewhere terms 'the conventional stage sets that are erected around us'[3] – revealing what this vision masks. But in what sense might the novels be considered to be bringing to light a *hidden* knowledge? If, say, they are concerned with the bodily and psychic risks of inhabiting a highly technologized world, then the sense of risk could be said to be one that many of this world's inhabitants consciously share. For the purposes of answering our question it is useful to take seriously Ballard's declared interest in the thought of Freud: the novels' reflections on the relationship between psyche and the ambient culture are a recognizable development out of a central psychoanalytic concern and form of enquiry.

In the classic Freudian account passage through the Oedipus complex brings the individual subject to the position where, for the first time, s/he finds that his/her wishes are aligned with the imperatives of the culture:

> The Oedipus complex is not reducible to an actual situation – to the actual influence exerted by the parental couple over the child. Its efficacy derives from the fact that it brings into play a proscriptive agency (the prohibition against incest) which bars the way to naturally sought satisfaction and forms an indissoluble link between *wish* and *law*.[4]

The individual is facing not (just) the empirical father or parent but cultural rules and norms in their immoveability. Imposition of this inflexible law is, however, experienced as a kind of empowerment. As Lacan ingeniously puts/puns it, the *nom du père* is also in an immediate conversion its homophone, the *non du père:* the prohibition is also a placing in the system where one finds as one's own a proper name and identity; a sense of unsustainable confrontation paradoxically mutates into a sense of freedom. From the oedipalized standpoint cultural law is on the side of the desiring subject; this happy view is sustained largely by the pressure of unresolved anxiety, now barred from consciousness except in this anxiety's unrecognized displacements.

In the novels' perspective, the recent-modern, relatively sudden extension of technology's power to shape human behaviour at levels ranging from the most intimate to the most public has a psychic significance directly comparable to that of oedipalization in the classic sense. We are dealing here primarily with the burgeoning technologies of medicine, communications and transport. They offer, in their differing though often crossing ways, something akin to a prosthetic extension of human capabilities (Freud in 1930 famously characterized modern, technologically augmented Man as 'a kind of prosthetic God'[5]): the injured are repaired, biological process is superseded, the effects of distance are commuted or abolished. As a condition of the prosthesis' acceptance, desires are fixed in the form that the prosthesis facilitates – a phenomenon particularly apparent in the case of the car and its cult, offering as personal enhancement exactly those things which the car allows one to do. But this prosthesis has also been encountered, in a psychic moment now lost to conscious knowledge, as alien, frightening and intrusive. In the account implied in the 'Crash!' section from *The Atrocity Exhibition* cars, being inherently 'violent and unstable machines' (author's commentary, p. 109), mobilize powerful sado-masochistic fantasies ('As part of a continuing therapy programme, patients devised the optimum wound profile [of crash victims]'; p. 111). The rationally disturbing fantasies outlined in this section of the novel support the commercial cult of the car in that it functions to hold them at bay, to keep their pressure latent: the car is

> the most powerfully advertised commercial product of this century, an iconic entity that combines the elements of speed, power, dream and freedom within a highly stylized format that defuses any fears we may have of the inherent dangers (author's commentary, p. 109).

Oedipalized behaviour here has a perverse subliminal underpinning, qualitatively different from conscious knowledge that cars can kill.

Trauma and its functional concealment perform a comparable role in the propagation of later 60s Reaganite politics as evoked in another non-narrative section of *The Atrocity Exhibition* ('Why I Want to Fuck Ronald Reagan', the presence of which caused the pulping of the book's first American edition; this piece was written, as Ballard notes, when the Governor of California was becoming established as a contender for the Presidency, and receiving a great deal of media attention). Ballard's marginal commentary recalls that Ronald Reagan, as film actor-turned-politician, was both an accomplished performer before the camera and a familiar one whose accomplishment audiences could accept without misgiving. Witnessing his television performances, they seem to have been primarily aware of and to have responded favourably to a certain blandness of self-presentation: 'In his commercials Reagan used the smooth teleprompter-perfect tones of the TV autosalesman.' Latent in this self-presentation was a more disturbing one: the content of Reagan's speeches, supposing that one paid attention to this, suggested 'a crude and ambitious figure, far closer to the brutal crime boss he played in the 1964 movie, *The Killers*' (author's commentary, p. 119). Within the text of the novel, experimental subjects 'in terminal paresis (G.P.I.)' supply the subliminal content of the presidential contender's political performance in the form of anal-sadistic fantasy:

> Subjects were required to construct the optimum auto-disaster victim by placing a replica of Reagan's head on the unretouched photographs of crash fatalities. In 82 percent of cases massive rear-end collisions were selected with a preference for expressed fecal matter and rectal haemorrhages. (p. 119)

Central to the performances' political success was, however, the discontinuity between levels of perception, enforced by the co-presence of traumatic content and surface blandness:

> Above all, it struck me that Reagan was the first politician to exploit the fact that his TV audience would not be listening too closely, if at all, to what he was saying, and indeed might well assume from his manner and presentation that he was saying the exact opposite of the words actually emerging from his mouth (author's commentary, pp. 119–20).

Here a practice of cultural oedipalization (the establishment of a regulatory structure as wish) mobilizes and requires the support of perverse fantasies which its structure at the same time witholds from conscious scrutiny.

In this account the propagation of Reaganite politics involves, like infantile passage through the Oedipus complex, the installation of a set of directing wishes, felt to empower, as the subject's own; the conversion of trauma into empowerment is, once again, attended by strong effects of lawlike-ness and binding necessity – successful oedipalization is an epiphany of Law. But the regulatory principle that emerges seems to require a footing in non-Oedipal psychic conditions which it also defines as perverse. In a Freudian understanding, pre-Oedipal fantasy is directed towards entities which its very structure posits as unattainable; in a central orientation, perfect fusion of self with (the breast of the) Mother. Oedipalized fantasy, on the other hand, has gained access, via prohibition, to the concretely envisaged *possibility* of attaining the object that is prohibited. As Slavoj Žižek explains, the myth of the primal father which Freud elaborates in *Totem and Taboo* (1912–13) thus supplements the Oedipus myth

> by embodying [what had been previously defined as] impossible enjoyment in the obscene figure of the Father-of-Enjoyment, i.e. in the very figure who assumes the role of the agent of prohibition. The illusion is that there was at least one subject (the primal father possessing all women) who was able to enjoy fully.[6]

The imaginary prohibitor and definer of law-like desires is thus, latently, the one who escapes the Law's operation (Lacan writes *'perversion'* as *'père-version'*, or turning towards the imaginary Father). Evocations of sexual obscenity and lawless brutality in Reagan's media self-presentation can thus be strangely complicit with the establishment of Reagan as a paternally reassuring, benevolent political leader.

Opening out for imaginative exploration what these sections of *The Atrocity Exhibition* primarily state, *Crash* frames a narrative which disrupts the circuit between an oedipalized cultural 'reality' and its veiled traumatic scene.[7] Entered via the narrator's ('Ballard's') initiatory car crash, this scene now becomes a quasi-theatrical one in which 'Ballard' can move freely between the roles of spectator and object of spectatorship. This second spectatorship is one that Vaughan and the camera image also offer 'Ballard' for the sharing – the narrative is in movement towards imaginary completeness of vision.[8] How does the first car crash yield a new form of vision? As 'Ballard' reflects during his period of hospitalized recovery, participation in a serious car crash has been vitalizing: 'the crash was the only real experience I had been through for years'. 'Real' here includes the sense of forced encounter with what would otherwise have remained unknown: 'for the first time I was in physical confrontation with my own body, an inexhaustible encyclopaedia of pains and discharges, with the hostile gaze of other people, and with the fact of the dead man'. But it also includes the experience of being made the almost entirely passive object of techno-scientific knowledge and ministration. What is knowable from this position must be that of which the culture's most prestigious agencies of knowledge already possess understanding, and which they have in some sense determined:

> After being bombarded endlessly by road-safety propaganda it was almost a relief to find myself in an actual accident. Like

> everyone else bludgeoned by these billboard harangues and television films of imaginary accidents, I had felt a vague sense of unease that the gruesome climax of my life was being rehearsed years in advance, and would take place on some highway or road junction known only to the makers of these films. At times I even speculated on the kind of traffic accident in which I would die.[9]

On the heel of these thoughts 'Ballard' is taken off to be X-rayed; it is as if he is witnessing the disappearance of an interiority that could be witheld from authoritative practices of knowledge and control.

A similar idea has been articulated in 'Ballard's' visual fascination with the young woman who confronted him in the other crashed car, and with the utter infringement of what had a few moments before been her secure privacy: 'Still wearing her seat belt, the dead man's wife was coming to her senses. A small group of people – a truck driver, an off-duty soldier in uniform and a woman ice-cream attendant – were pressing their hands at her through windows, apparently touching parts of her body' (p. 21). For 'Ballard' as for this other survivor, the experience of potentially unlimited intrusion melds with and tends to replace that of the crash as physical trauma. Immobilized behind the wheel of his crashed car, 'Ballard' (in this fiction one who professionally deals with the filmic image) has become a knowing participant in the culture's virtual event. He is an adult suddenly infantilized, in that what he is and will become is now largely subjected to the determination of medicine and techno-science, and that the determination can be seen as lawless:

> For a moment I felt that we were the principal actors at the climax of some grim drama in an unrehearsed theatre of technology, involving these crushed machines, the dead man destroyed in their collision, and the hundreds of drivers waiting beside the stage with their headlamps blazing. [...] An uneasy euphoria carried me to the hospital. I vomited across the steering wheel, half-conscious of a series of unpleasant fantasies. Two firemen cut the door from its hinges. [...] Even their smallest movements seemed to be formalized, hands reaching towards me in a series of coded gestures. If one of them had unbuttoned his coarse serge trousers to reveal his genitalia, and pressed his penis into the

> bloody crotch of my armpit, even this bizarre act would have been acceptable in terms of the stylization of violence and rescue (pp. 22–3).

The authority that prevails here, with what is felt to be massive spectatorial assent, is at once highly codified and unpredictable, anarchic and binding, in what it exacts of those in its power.

Crash-enforced encounter with what is secretly deranged in the ordinary functioning of cultural authority becomes the novel's threshold to two quite different kinds of apocalyptic vision. Through the encounter with Vaughan, 'Ballard' and the narrative gain imaginative access to an affectively drained, desolate world in which human beings have already undergone the ultimate disaster. Driving 'Ballard' from the Road Research Laboratory while 'Ballard' scans the appalling results of his scientific researches, Vaughan is seen thus:

> Vaughan leaned against the window-sill, fingers raised to his nostrils as if clinging to the last odour of semen on their tips. The warning headlamps of the oncoming traffic, and the overhead lights of the expressway, the emblematic signals and destinations, lit up the isolated face of this hunted man at the wheel of his dusty car. I looked out at the drivers of the cars alongside us, visualizing their lives in the terms Vaughan had defined for them. For Vaughan they were already dead. (p. 137)

Vaughan as 'hoodlum scientist' (p. 19) responds to and projects the imperatives of techno-scientific culture in their incoherent, senseless aspect. The world that his activities posit has many of the characteristics of Baudrillardian 'hyper-reality', whose principle is that the real is entirely available as model, securing the possibility of its indefinite reproduction by specifiable means (in the text cited at the head of this article Baudrillard responds enthusiastically to those features of *Crash* that can be imaginatively organized around the figure and envisaged project of Vaughan). Baudrillard characterizes 'hyper-reality' as follows:

> [The real] no longer has to be rational, since it is no longer measured against some ideal or negative instance. It is nothing more than operational. [...] It is no longer a question of imitation, nor of reduplication, nor even of parody. It is rather a question of

> substituting signs of the real for the real itself. [...] Never again will the real have to be produced: this is the vital function of the model in a system of death.[10]

'Ballard's' narration presents a Vaughan who acts in permutative relation to elements of car design and conditions of driving, as for example in the scene with the young prostitute which is styled a 'marriage of sex and technology' (pp. 142–5). In the pursuit of his own sexual-scientific researches Vaughan experiments with decontextualized motifs of violence and desire, searching for the optimal combination; these motifs are often direct imprints of human bodily reality, in a series which includes scrawled photographs and pieces of fabric bearing human stains and secretions, now valued purely in their character of signs.

Vaughan is drawn to the Road Research Laboratory by the efficacy of its simulations. Observing its engineers prepare their test car, he explains to 'Ballard' that 'Using this set-up they could duplicate the Mansfield and Camus crashes – even Kennedy's – indefinitely' (p. 123). What 'Ballard' sees at the Laboratory, and generally, differs however from what Vaughan sees; this despite the importance of Vaughan in supplying components of a hitherto-unfamiliar landscape of vision. Viewing the laboratory-staged crash presented once again as filmed image, 'Ballard's' awareness is directed not to possibilities of repetition but to those of transfiguration and the releasing of concealed possibilities:

> In a dream-like calm, the front wheel of the motorcycle struck the fender of the car. As the rim collapsed, the tyre sprung inwards on itself to form a figure of eight. The tail of the machine rose into the air. The mannequin, Elvis, lifted himself from his seat, his ungainly body at last blessed by the grace of the slow-motion camera. Like the most brilliant of all stunt men, he stood on his pedals, legs and arms fully stretched. His head was raised with its chin forwards in a pose of almost aristocratic disdain. The rear wheel of the motorcycle lifted into the air behind him, and seemed about to strike him in the small of the back, but with great finesse the rider detached his feet from the pedals and inclined his floating body in a horizontal posture. (p. 126).

Confronted with the derangement of cultural authority, 'Ballard', unlike Vaughan, searches for a hitherto-concealed arrangement, in a mutation of consciousness that transforms Vaughan from hoodlum scientist to the 'nightmare angel' (p. 84) of a revelatory vision. The vision is informed by a sense of impending catastrophe, personal and collective (the cars gridlocked on the expressways surrounding 'Ballard's' and Catherine's apartment seem, for example, to be rehearsing the 'coming autogeddon'; p. 50), but also of the latency of a new world in a superseded one. Shaped by this double expectancy, the 'Ballard'-centered narrative of *Crash* significantly reabsorbs and recontextualizes a series of motifs deriving from the mainstream of Western apocalyptic writing.

'Ballard' accepts, at first as personal insight and then as confirmed by the encounter with Vaughan, a massive simplification of consciousness in the interests of comprehensive seeing; the authority of the car and of attendant technologies now dictates the fundamental form of what can be envisaged. Returning to his apartment in West London after the crash, 'Ballard' notices that the presented field of vision is now literally enframed by roadways:[11]

> During my weeks in hospital the highway engineers had pushed [the] huge decks [of the airport-access motorway] more than half a mile further south. Looking closely at this silent terrain, I realized that the entire zone which defined the landscape of my life was now bounded by a continuous artificial horizon, formed by the raised parapets and embankments of the motorways and their access roads and interchanges. These encircled the vehicles below like the walls of a crater several miles in diameter. (p. 53)

Acts of perception have already undergone a figurative enframing; the gaze is directed in such a way that a familiar, dispersed sense of reality recedes while a more urgent and coherent one takes its place. In the reflections that precede this passage 'Ballard' describes a new form of seeing whose scope and orientation are supplied by a monumental technology serving the car and its needs:

> I realized that the human inhabitants of this technological landscape no longer provided its sharpest pointers, its keys to the borderzones of identity. [...] All the hopes and fancies of this placid suburban enclave, drenched in a thousand infidelities, faltered before the solid reality of the motorway embankments, with their constant and unswerving geometry, and before the finite areas of the car-park aprons. (pp. 48–9)

Shortly afterwards Vaughan, strangely calm and assuming the manner of 'an instructor ready to help a young pupil', will explain to 'Ballard' via his telescopic photographs that the central relationships of 'Ballard's' current life are 'mediated by the automobile and its technological landscape' (pp. 102, 101). This is a teaching amusingly recapitulated by the images in Vaughan's Danish sex magazines 'handed to me, presumably as a pacifier', in all of which 'the motor-car in one style or another figured as the centrepiece' (pp. 101–4). Another of the novel's figures for the contraction of consciousness takes the form of preoccupation with the internal space and layout of cars, as distinct from cars as bodies in movement: in an early stirring of the idea 'Ballard's' wrecked car in the police pound, encoding the memories of his recent sexual involvements, itself becomes a 'small museum of excitement and possibility' (p. 69).

This contraction or obsessive centering of consciousness is also, however, the precondition for the narrative's expansive, connective dealings with the scene of trauma. This scene undergoes massive and literal extension in 'Ballard's' car-borne physical movements, establishing a kind of equivalence between the location of the original crash and other nodes of excitement and pleasure. At first 'Ballard' makes a series of half-involuntary returns to this first location, broken in increasingly for pleasurable response, manifestly dangerous since the returns are the crash's near-restaging (here he is in the company of the crash's other living participant):

> We sped towards the junction with the Drayton Park motorway spur. She [Helen Remington] steadied herself against the chromium pillar of the quarter-window, almost dropping her cigarette on to her lap. Trying to control the car, I pressed the head of my penis against the lower rim of the steering wheel. The car swept towards its first impact point with the central reservation. [...] Semen jolted through my penis. As I lost control of the car the front wheel struck the kerb of the central reservation, throwing a tornado of dust and cigarette packs on to the windshield (p. 74).

But 'Ballard's' repetitive journeys on the roadways of West London also define the topography of a larger scene, which in terms of psychic need replaces and extends the meaning of the crash location. Many of these journeys lead to or yield sight of the distinctive non-natural terrain of Heathrow airport, a mandala-like enclosure within an enclosure of roads and place of contact with the imagination of vertical movement. The novel's visibly-bounded, technologically determined landscapes, which have the airport as their exemplar and co-ordinating centre, offer themselves to 'Ballard' as containing the complete possibilities of experience – a motif half-mockingly presented in the three airport prostitutes who 'seemed to form a basic sexual unit, able in one way or another to satisfy all comers' (p. 61).

Crash at its most visionary-apocalyptic is the discovery, via the trauma of bodily and psychic catastrophe, of a new city, with new inhabitants, which has latently existed in the old city, now superseded. Low-key descriptive passages like the following play a complex role in the process of discovery:

> We left the overpass and moved down a concrete road through west Northolt, a residential suburb of the airport. Single-storey houses stood in small gardens separated by wire fences. The area was inhabited by junior airline personnel, car-park attendants, waitresses and ex-stewardesses. Many of them were shift-workers, sleeping through the afternoon and evening, and the windows were curtained as we wheeled through the empty streets (p. 87).

The passage takes precise measure of a very ordinary suburban landscape whose routines and unemphatic social distictions are produced by the needs of air travel; 'Ballard' as recipient of a revelation for which no sense of the ordinary would prepare him is also glimpsing a kind of temple-complex dedicated to the extravagant religion of flight. Having returned Seagrave, the damaged stunt-driver, to his home in a differently defined marginal area of 'breakers' yards and vehicle dumps, small auto-repair shops and panel beaters' (p. 92), 'Ballard' will find himself in a party of those bonded by their psychic and physical relation to the phenomenon of the car crash. He sees Vaughan's photographs of Gabrielle, who has already attracted his attention at the party, as charting the process by which an 'agreeable young woman, with her pleasant sexual dreams, [has] been reborn within the breaking contours of her crushed sports car'; the process's imputed effect is that she can find in the 'deformed panels' and 'twisted instrument binnacles' of her new invalid car 'a readily accessible anthology of depraved acts, the keys to an alternative sexuality' (pp. 99–100).

The novel's world already contains a visually perfect body – that of 'Ballard's' wife Catherine, the 'forgery of an Ingres' ('What had first struck me about Catherine was her immaculate cleanliness, as if she had individually reamed out every square centimetre of her elegant body'; p. 112). The narrative's apocalyptic movement is, however, towards replacement of the ordinarily desired body with one whose form and sexuality are redefined by the violent impact of a machine's 'breaking contours'. At an intermediate stage in this movement 'Ballard' discovers a special tenderness in his relation to the body of Catherine as lightly deformed by the beating that she has received from Vaughan: 'She watched me with a calm and affectionate gaze as I touched her body with the head of my penis, marking out the contact points of the imaginary automobile accidents which Vaughan had placed on her body' (p. 166). Vaughan's carefully chosen medical photographs show how directly the body can be reconfigured by impact with a car's internal surfaces and projections (the following passage is apparently based on a real textbook, Jacob Kolowski's *Crash Injuries: The Integrated Medical Aspects of Automobile Injuries and Deaths*):[12]

> In several of the photographs the source of the wound was indicated by a detail of that portion of the car which had caused the injury: beside a casualty ward photograph of a bifurcated penis was an inset of a handbrake unit; above a close-up of a massively bruised vulva was a steering-wheel boss and its manufacturer's medallion. [...] [I]n the photographs of facial injuries [the] wounds were illuminated like medieval manuscripts with the inset details of instrument trim and horn bosses, rear-view mirrors and dashboard dials. The face of a man whose nose had been crushed lay side by side with a chromium model-year emblem (pp. 134–5).

In fulfilment of the photographs' revelatory meaning 'Ballard' arrives through sexual encounter with the crash-transformed Gabrielle at a reconfigured sexuality, prizing as erogenous exactly those physical formations which collision with specifiable parts of a machine has produced (he 'celebrate[s] with her the excitements of those abstract vents let into her body by sections of her own automobile'; p. 179). Anal penetration of Vaughan, as envisaged when he arrives in 'Ballard's' office with its 'enlarged sales photographs of automobile radiator grilles and windshield assemblies' (p. 147), has been an alluring matter of enclosure within and bodily contact with a car's lucidly imagined body ('I vizualized these sections of radiator grilles and instrument panels coalescing around Vaughan and myself, embracing us as I pulled the belt from its buckle and eased down his jeans, celebrating in the penetration of his rectum the most beautiful contours of a rear-fender assembly'; p. 148). In both cases the sexual relation seems to have become a relation with the utterly known and completely visible as modelled by the elements of car design.

The novel's world of neo-apocalyptic revelation is finally a world saturated with knowledge. In such a world loss is impossible, because the very form of loss is known – loss becomes transformation into a cognitively complete entity. While embracing Gabrielle 'Ballard' visualizes 'as Vaughan had taught me [...] extraordinary sexual acts celebrating the possibilities of unimagined technologies'. In one of these visualizations Catherine has had her face destroyed in a crash where the 'splintering steering column' has also opened 'a new and exciting orifice in her perineum, [...] neither vagina nor rectum'. (Catherine's untransformed body includes, conversely, armpits which are 'tender fosses like mysterious universes'.) In another

> I visualized the body of my own mother, at various stages of her life, injured in a succession of accidents, fitted with orifices of ever greater abstraction and ingenuity, so that my incest with her might become more and more cerebral, allowing me at last to come to terms with her embraces and postures. (pp. 179–80)

Road vehicles as they appear in 'Ballard's' climactic acid vision become the bearers of near-limitless information, while at the same time transcending their ordinary physical conditions of movement in a scene which explicitly melds roadway with airport:

> An armada of angelic creatures, each surrounded by an immense corona of light, was landing on the motorway on either side of us, sweeping down in opposite directions. They soared past, a few feet above the ground, landing everywhere on these endless runways that covered the landscape. I realized that all these roads and expressways had been built by us unknowingly for their reception. (p. 199)

Later, when the acid trip has itself become a trauma requiring a period of recovery, 'Ballard' will scan the landscape below his flat 'trying to find this paradisal incline, a mile-wide gradient supported on the shoulders of two angelic figures, on to which all the traffic in the world might flow' (p. 208). The visionary landscape that he now attempts to recompose possessed, like the new city which descends from heaven at the end of the book of Revelation, the properties of maximally visible organization in three dimensions, centredness, and potential all-inclusiveness.

Vaughan is the angel-like donor and sustainer of this vision (his relation to 'Ballard', nearly incapable of driving, is that of 'parent' to 'exhausted child'; p. 199), as well as the provider of the LSD; he is not, however, fully susceptible of being included within it, since he remains the narrative's marker of the indissoluble linkage between vision and trauma – in this perspective trauma might be said to *produce* vision as its imaginary overcoming. When 'Ballard' seduces Vaughan in the car which 'glow[s] like a magician's bower', he perceives him as 'a wholly benevolent partner, the eye of this illumination of the landscape around us' (pp. 199–200). But 'Ballard' also, for a moment, finds himself grappling with a stranger being (an angel of a different kind if the reference is to Jacob's struggle by the River Jabbok):[13]

> I held his face in my hands, feeling the porcelain smoothness of his cheeks, and touched with my fingers the scars on his lips and cheeks. Vaughan's skin seemed to be covered with scales of metallic gold as the points of sweat on his arms and neck fired my eyes. I hesitated at finding myself wrestling with this ugly golden creature, made beautiful by its scars and wounds (p. 201).

Scars and wounds are the intractable physical marks of trauma. To 'Ballard' in an earlier encounter Vaughan's scars have been imagined as bearing precise information: 'These apparently meaningless notches on his skin [were in fact] a cuneiform of the flesh [which, together with the car components which could be imagined as having produced them,] described an exact language of pain and sensation, eroticism and desire' (p. 90). Vaughan, on the other hand, picks at his scars, which bleed once again, and augments them through voluntary wounding (see pp. 191–2). It seems characteristic of *Crash* that it should draw a circle of completed meaning, and at the same time set against the perfection of this figure a brutal disarray of marks let into the body.

Notes

1. Jean Baudrillard, 'Ballard's *Crash*', trans. Arthur B. Evans, *Science-Fiction Studies*, 18 (1991), p. 314.
2. *The Atrocity Exhibition* with author's annotations added from the edition of 1990; see note 8 (London: Harper Collins, 1993), p. 111. Cf. the comments on the same Freudian distinction in the author's Introduction to *Crash* (London: Vintage, 1995), p. 5.
3. 'Interview by Greame Revell', in *Re/Search*, nos. 8/9 (1984), p. 47.
4. J. Laplanche and J. B. Pontalis, *The Language of Psycho-Analysis*, trans. Donald Nicholson-Smith (London: The Hogarth Press, 1983), p. 286.
5. Sigmund Freud, *Civilization and its Discontents*, Penguin Freud Library, vol. 12 (Harmondsworth: Penguin Books, 1985), p. 280.
6. Slavoj Zižek, *Looking Awry: An Introduction to Jacques Lacan through Popular Culture* (Cambridge, Mass.; MIT, 1992), p. 24.
7. For a differently centred account of the role of trauma and the nature of narratorial response to it in *Crash* and other novels of Ballard, see Dennis A. Foster, 'J. G. Ballard's Empire of the Senses: Perversion and the Failure of Authority', *PMLA* 108 (1993), pp. 519–32. Foster, also drawing on Lacan, writes of a return in fantasy to a pre-Oedipal sense of the body as corporeally fragmented or entirely available to knowledge as surface. I am indebted to Foster's discussion, but my own stresses what is perverse in the ordinary processes of oedipalization – from this standpoint, cultural authority is always already a scandalous failure.
8. Cf. in the 1990 edition of *The Atrocity Exhibition* which was revised and supervised by the author (San Francisco: Re/Search Publications), the anatomical, exoscopic drawings of Phoebe Gloeckner, intercalated with photographs by Ana Barrado which emphasize hard, gleaming surfaces.
9. *Crash* (London: Vintage, 1995), p. 39.
10. Jean Baudrillard, 'Simulacra and Simulations', in *Selected Writings*, ed. Mark Poster (Oxford: Blackwell, 1988), p. 167.
11. The carefully tracked emergence of a framing (going with centering and simplification) of narratorial consciousness differentiates the narrative method of *Crash* from those of the more parable-like *Concrete Island* (1974) and *High-Rise* (1975), where the main scene of action is already enframed.
12. Published Springfield: Thomas, 1960; see Foster, 'J. G. Ballard's Empire of the Senses', p. 524.
13. Genesis 32: 22–32; in traditional interpretation of the passage, Jacob's mysterious opponent is an angel.

11

Disguising Doom: A Study of the Linguistic Features of Audience Manipulation in Michael Moorcock's *The Eternal Champion*

MICHAEL HOEY

INTRODUCTION

Michael Moorcock's *The Eternal Champion* (1970) has a simple enough plot. Humans on a parallel Earth call to their aid an eternal champion, Erekosë, to help them rid the world of the Evil Ones, the Eldren. He does so, almost wiping them out. At the last he realises that they were not evil after all but that the humans were the truly evil ones. So he switches sides and proceeds to wipe out the human race:

> Two months before I had been responsible for winning the cities of Mernadin for Humanity. Now I reclaimed them in the name of the Eldren [...] I destroyed every human being occupying them [...] Not merely the great cities were destroyed. Villages were destroyed. Hamlets were destroyed. Towns and farms were destroyed. I found some people hiding in caves. The caves were destroyed. I destroyed forests where they might flee. I destroyed stones that they might creep under [...] It was fated that Humanity should die on this planet.[1]

And there the story ends, with the human race extinct and the Eldren reduced to a tiny remnant. It is not hard to see it as a parable of apocalypse reflecting the nuclear war fears of the period. The book was first published in its full form in 1970, though part of it was published as early as 1956, and Erekosë wields a radioactive sword on a parallel Earth in 'the twentieth century AD in the Age of Men' (p. 7). The demonisation of the communist world by the West (and vice versa) at that time is reflected in the demonisation of the Eldren, and the notion of an enormous power of destruction that ends up annihilating the people who invoke it has its parallels in numerous popular apocalyptic fictions of the sixties (perhaps most notably *Dr. Strangelove*). My interest in this book, however, lies elsewhere, namely in how Moorcock succeeds in portraying a hero who is unremittingly destructive and commits two great acts of genocide and yet does not turn into a villain. Related to this is another question: how does he portray the human race such that we do not see them as evil at the outset of the novel and yet are able to see them as evil at the end?

Jacqueline Harvey shows how Arthur C. Clarke, Poul Anderson and Ray Bradbury, amongst others, use sleight of word in order to disguise the direction and intent of their stories.[2] Sometimes the device is one of using a common expression with a rarer meaning, sometimes it is a matter of allowing a reader to draw incorrect inferences from technically correct statements, sometimes it is just a question of permitting readers to make generic assumptions on the basis of apparently adequate but actually inadequate information. Her work provides a clue as to where we might find possible answers to the questions I have just asked – in Moorcock's manipulation of lexical and text-linguistic features in the early part of the text.

Almost all of my attention will be given to the prologue of the book and to a couple of passages from the first chapter. It is here, I shall argue, that all the subsequent stages of the narrative are prepared for, both the assumption that the humans are on the side of the good and the carefully concealed clues that they may not be. Likewise, I shall argue, the hero is portrayed as (or portrays himself as) victim in the very first pages so as to reduce the chance of our blaming him for the apocalyptic atrocities that follow.

The first words of the Prologue and consequently the book are as follows:

> *They called for me. That is all I really know. They called for me and I went to them. I could not do otherwise. The will of the whole of humanity was a strong thing ... Why was I chosen? I still do not know, though they thought they had told me.* (p. 8) [italicised in the original]

Chapter 1, 'A Call Across Time', reveals that that the call described in the first three sentences of the Prologue is a physically real one:

> Then, between wakefulness and sleeping, I began every night to hear voices [...] At first I dismissed them, expecting to fall immediately asleep, but they continued, and I began to try to listen to them, thinking, perhaps, to receive some message from my unconscious [...] I could not recognise the language, though it had a peculiar familiarity. (p. 9)

However it seemed to me that there was an oddity about the first four words of the Prologue that was not apparent on first reading, an oddity compounded in the third sentence whose first half repeats the first sentence in its entirety. As a way of investigating the clause's apparent oddity, I decided to look at some of the expressions of calling current in the language.

A corpus-linguistic approach to the study of any linguistic phenomenon requires that one adopts a corpus of relevant material and elicits from that corpus as many instances as possible of the linguistic phenomenon under consideration. The resultant list of instances is referred to as a concordance, and on this concordance one can then perform various kinds of computational and statistical operation in order to discover the (often hidden) regularities within it. These regularities may be lexical, grammatical, or semantic. With regard to expressions of calling, I examined a corpus of 4½ million words made up of a variety of genres (including fiction), drawn in part from the British National Corpus and in part from a corpus of *Guardian* editorial features, and created concordances of the three phrases *x called y*, *x called to y* and *x called for y*, where *x* and *y* stand for noun phrases or pronouns, making use of WordSmith's sophisticated software for the purpose.[3] WordSmith permitted me to look at all instances of these expressions in their larger contexts and to sort them in any way I wanted.

In my 4½ million word corpus there were 41 examples of *x called y* (where *called* was used as an active verb not as a past participle). Of these, over two-thirds (28) had the meaning of summoning, either by phone or more directly, for example:

> On one occasion he called a player into his office, took his hat from the hat-stand and threw it on the floor.

> After taking an overdose he told his neighbour what he had done and she called an ambulance.

The verb *called* would therefore have conveyed the sense of summoning apparent in the prologue and first chapter. The verb phrase *called to* would however have done the job even better. I found 31 instances of *x called to y* in my corpus, excluding idiomatic expressions such as *I called to mind.* All but two of these were found to have the meaning of shouting to someone; at least fifteen had the meaning of summoning from a distance which is the particular meaning that Moorcock is apparently conveying, for example:

> As I waited, Ryan came out of a near-by block and called to me: 'Can I have a word with you, old boy?'
>
> Here I had to find some kind of billet, to the rear of the trench, in a flimsy shelter under a corrugated roof covered with a few sandbags; five men were lying side by side. One that I knew called to me: 'Come in, we can make room for you.'

In the following example from Genesis, in the New International Version translation, we read:

> God heard the boy crying, and the angel of God called to Hagar from heaven and said to her, 'What is the matter, Hagar?'[4]

Distances do not come any greater.

So either *called* or *called to* would have served Moorcock's ostensible purpose well, and indeed *called to* would have apparently served his purpose exactly: Erekosë is being summoned across a great distance and across time. The picture for *called for,* the expression Moorcock actually used, is however markedly different. Of the 42 examples of *x called for y* recovered from the four-and-a-half-million-word corpus, only five were found to be used with any sense of summoning and two of these were instances of calling for beer! So the oddity in the first four words of the book lies in the fact that the normal expression for the physical call from one person to another across some distance is *called to* or simply *called*, not *called for*.

Why would Moorcock apparently use the wrong collocation with which to begin his book? An answer perhaps lies in the fact that of the 42 examples of *x called for y* in my corpus 29 (69 per cent) refer to the solving of some problem, the improvement of some situation or the rectification of some injustice, such as:

> Why were we silent when General Morillon called for military action to stop the killing in Srebenica?
>
> No sooner had it been evacuated than the Labour MP for the constituency called for a public inquiry.

Thus the first four words of the book conjure up the idea that Erekosë is the hoped-for source of a solution to humanity's problem, or a way in which humanity can rectify an injustice or simply improve its situation. From the very first sentence, in other words, a problem–solution pattern is hinted at with Erekosë in the hoped-for Response slot, of which we will have more to say in a moment.

The next point to note about this sentence is that what is *called for* is often help. It is impossible to calculate exact collocations on the basis of 42 examples, useful though such a sample is from the point of view of examining general semantic patterns. I therefore examined 12166 examples of *call* for*,[5] found in a corpus of approximately 100 million words, mainly drawn from the *Guardian*. I found that *help* was the sixth most common word to appear immediately after *for*. *Aid* was the fiftieth most common word in the position two words to the right of *for* (as in *called for renewed aid*). So Erekosë is a potential source of help/aid, and this, of course, in turn means that he is in Propp's terms a hero or a donor.[6] So the very first words type him in such a way that even his being the direct cause of an apocalypse cannot affect the characterisation.

But it is not only Erekosë who is typed by the first few words. The sentence also types the callers: *they* are encoded as having done the calling. They are therefore in the position of having identified a problem and of seeking help. To see what kind of people characteristically in the language are described as calling for help, I re-examined the data and in particular the 28 cases of *called for* that were associated with problem-solving and so on in the original search. I found that all but four were associated with leaders, representatives (or laws passed by these) or ginger groups seeking to make life better for some other group of people.

In the first four words of the book, therefore, we have the space created for a hero and leaders calling for that hero, a pattern that goes back to *Beowulf*. We do not learn for another 64 sentences, well into the first chapter, that the callers are in fact the King and his daughter, the leaders of all humanity. They fit well, then, into the regular pattern in the language that *callers for* are people of authority. Importantly, the fact of their being *callers for x* is not only confirmation of their authority but confirmation simultaneously of their altruism in so doing. People who *call for* solutions to problems are rarely trying to solve their own problem.

So here we have the first step towards answering to the questions we were considering earlier: humanity is in the very first words depicted as having a problem, which the leaders are seeking to solve by calling for help to someone who will be a response to the problem. It is hard (though importantly not impossible) to equate this with an evil humanity led by paranoid leaders who call up an immensely destructive force for whom power is right.

One of the most common structures of narrative is the problem–solution pattern,[7] a classic version of which takes the form:

Problem – Response – Positive Evaluation/Positive Result

Each of these narrative stages is characteristically attributed to some participant (or to the narrator or reader), as illustrated in the following narrative:

When the water in the skin was gone, she [Hagar] put the boy under one of the bushes. Then she went off and sat down nearby, about a bow-shot away, for she thought, 'I cannot watch the boy die'. And as she sat there nearby, she began to sob.	Problem for Hagar
God heard the boy crying, and the angel of God called to Hagar from heaven and said to her, 'What is the matter, Hagar? Do not be afraid; God has heard the boy crying as he lies there. Lift the boy up and take him by the hand, for I will make him into a great nation.' Then God opened her eyes and she saw a well of water. So she went and filled the skin with water and gave the boy a drink.	Response by God
God was with the boy as he grew up.[8]	Positive Evaluation by narrator

The narrative just given represents the problem–solution pattern in its classic simplicity. Notice that Hagar is assumed to be on the side of the angels even before an angel appears to her. It is not beyond the wit of a writer to undermine such an assumption, but it is our default assumption nevertheless. In the opening moments of *The Eternal Champion*, then, the verb phrase *call for* has established that we have a Problem for 'them' (later identified as King of humanity and his daughter) and a Response by Erekosë and that 'they' are likely to have our sympathy and Erekosë our admiration. I am simplifying here, but not hugely so.

At the other end of the problem–solution pattern, once one encounters a positive Result and/or Evaluation, the pattern comes to a full stop. Thus in the example above we feel no need for further narrative content. If any more were supplied, it would belong to a new narrative with the same central character; it would not be a continuation of this narrative pattern. If we have a drastically negative Result and/or Evaluation, the effect is the same: there is no expectation of continuation (unless, of course, there are other participants). Consider the following joke told by my son when he was 12:

> There were three men who were called Fred, Bill and Joe, and they were caught by a firing squad and they were all going to be shot. So Fred was brought out and all the firing squad lined up ready to shoot him. As they were about to press their triggers, he cried 'Tornado!' And the firing squad all ran off thinking there was a tornado and he escaped. Then Bill was brought out. As they were about to press their triggers, he cried 'Hurricane!' And the firing squad all ran off thinking there was a hurricane, and *he* escaped. Last of all Joe was brought on. As the firing squad lined up he shouted 'Fire!' And they shot him.

The positive evaluation in the first two cases and the drastically negative result in the third bring each of the narrative patterns to an end.

In longer narratives this rarely happens immediately. Instead we get recycling. When a result and/or evaluation is negative but not irredeemably so, it has the effect of triggering either a new Problem or a further Response to the original Problem. Thus the traditional tale of the Enormous Turnip has repeated attempts to pull up the turnip fail; each failure triggers a new Response, taking the form of an additional participant to help with the pulling, and the pattern recycles as many times as the listening child can bear it until finally there is a successful Result with the turnip coming out of the ground. This, and the second positive Result of all the participants eating the fruit (or vegetable) of their labours, bring the pattern decisively to an end.

The problem–solution pattern set up in the first sentence of *The Eternal Champion* manages to thwart all these expectations. In the first place, Positive Results are meant to bring the pattern to an end, but in this case they do not. The Result of Erekosë's Response to Humanity's Problem, like that of the Response of God to Hagar's Problem or that of the Responses of Fred and Bill to their own Problem, is apparently entirely positive, in the sense that it completely and satisfactorily removes Humanity's Problem:

> A year of pain and death, and everywhere that the banners of Humanity met the standards of the Eldren, the basilisk standards were torn down and trampled.
>
> We put all we found to the sword ... I did not care what they called me – Reaver, Blood-letter, Berserker – for my dreams no longer plagued me and my ultimate goal came closer and closer.
>
> Until it was the last fortress of the Eldren left undefeated. Then I dragged my armies behind me, as if by a rope. I dragged them towards the principal city of Mernadin, by the Plains of Melting Ice. Arjavh's capital – Loos Ptokai.
>
> And at last we saw its looming towers silhouetted against a red evening sky. Of marble and black granite, it rose mighty and seemingly invulnerable above us. But I knew we should take it.
>
> (*The Eternal Champion*, pp. 130–2)

Yet the language in which the Positive Result is couched would make most readers pause – *a year of pain and death* – and within a few pages, despite the absoluteness of the victory, Erekosë has switched sides.

This leads to the second respect in which expectations set up at the beginning of the book are thwarted. The Eldren are all slain, their territory is conquered, they are down to one city. According to the conventions of the pattern, they have no chance of recycling their problem–solution pattern. And yet, with Erekosë's switch of allegiance, they are able to respond to their Problem and, indeed, in some respects end up with their Problem solved – Humanity is wiped out. What distinguishes this recycling from normal pattern recycling is that we are well into the novel before we are encouraged to see that the real pattern of the novel is not:

Problem for Humanity	= they fear the Eldren
Response by Humanity	= to use Erekosë against the Eldren
Positive Result for Humanity	= the Eldren are virtually all destroyed

but:

Problem for Humanity	= they fear the Eldren	
Response by Humanity	= they use Erekosë against the Eldren	= Problem for the Eldren
	the Eldren put up a struggle	= Response by the Eldren
Positive Result for Humanity	= the Eldren are virtually all destroyed	= Negative Result for the Eldren
		= Problem for the Eldren
	Erekosë is persuaded to change sides	= Response by the Erekosë
Negative Result for Humanity	= Humanity is is wiped out	= Positive Result for the Eldren

The point being made here is that Moorcock's novel is uncharacteristic in its use of the conventional problem–solution patterns, in that we do not normally expect the patterning to involve our switching sides mid-pattern even when reason might demand that we do so. We may develop a concern for the other participants in the pattern, but we do not characteristically lose our original identification with those for whom the pattern has originally been set up. In Roger Hargreaves' children's book *Mr Nosey*, for example, the people of Tiddletown attack Mr Nosey's nose with paint, pegs, hammer and saw, but we never question their right to do so because his nosiness has been articulated by Hargreaves as a Problem. We acquire sympathy for him during the course of his misadventures and want him to solve his own Problem of being the victim of brutal assaults on his nose, but we do not expect his Response to involve his becoming a greater Problem for the people of Tiddletown. In *The Eternal Champion*, however, this reversal of sympathy is what is demanded of the reader.

So this leads to our second question: how does Moorcock prepare for this reversal? The first mention of Humanity's Problem comes two pages into the first chapter with the first full speech Erekosë hears as he is dragged through time to twentieth-century humanity:

> *Erekosë. It is I – King Rigenos, Defender of Humanity …*
>
> *You are needed again, Erekosë. The Hounds of Evil rule a third of the world and humankind is weary with the war against them. Come to us, Erekosë. Lead us to victory. From the Plains of Melting Ice to the Mountains of Sorrow they have set up their corrupt standard and I fear they will advance yet further into our territories.*
>
> *Come to us, Erekosë. Lead us to victory. Come to us Erekosë. Lead us …*
>
> (*The Eternal Champion*, p. 10; italics in the original)

There are all the signals of a problem–solution pattern. We have the negative evaluations *Evil* and *corrupt*, we have the markers of attempted Response *Defender* and *war against them*, and we have the signals of negative Result of the attempted Response – *rule a third of the world* (though of course that means that Humanity rules two-thirds) and *they will advance yet further*, with the implication that they have already advanced far. In this context, the call to Erekosë becomes a further Response to the original Problem.

There is nothing in all this that alters the perception established in the first sentence of the prologue that King Rigenos is on the side of good. His daughter's reaction underscores this perception:

> *Father. This is only an empty tomb. Not even the mummy of Erekosë remains. It became drifting dust long ago. Let us leave and return to Necranal to marshal the living peers.*
>
> (*The Eternal Champion*, p. 11; italics in the original)

Firstly, this offers a Negative Evaluation of the attempted Response by her father of summoning Erekosë and articulates the real world assumption that it is a desperate throw, offering a more realistic alternative Response. In so doing, her words reinforce the assumption that the Problem is real for them.

However, when Erekosë first speaks of their Problem, there is the first slight clue of the later reversal of perception:

> This I knew to be the domain of the Eldren, whom King Rigenos had called the Hounds of Evil.
>
> (*The Eternal Champion*, p. 11)

Here Erekosë, who is supposedly writing/speaking from a point in time after the end of the novel, does not assert that the Eldren are Evil but only that King Rigenos asserts it.[9] Yet the avoidance strategy does not draw attention to itself because the expression Rigenos uses is metaphorical. It would not be expected that a speaker would assert someone else's metaphor. Compare the original above with the adapted versions below:

> This I knew to be the domain of the Eldren, whom King Rigenos had called Evil.
>
> (*The Eternal Champion*, p. 11 – adapted)
>
> This I knew to be the domain of the Eldren, who were the Hounds of Evil.
>
> (*The Eternal Champion*, p. 11 – adapted)

The first, I suggest, would have given the game away. The failure to aver Rigenos's evaluation would have led the reader to doubt whether the Eldren were universally thought of as Evil. The second, on the other hand, is odd, in that it suggests that an alternative nomenclature for the Eldren is that of the Hounds of Evil. The metaphor is suppressed in this reading. Even so, it would be hard to read it as other than an assertion by Erekosë of the evilness of the Eldren. Had Moorcock written this, not only would there have been no hint of the reversal to come but his narrator could have been accused of bad faith in saying that which he knew to be false.

This pattern of cunningly disguised refusal to aver Humankind's assessment of the Eldren as evil repeats itself a page later:

> The warrior kings of whom King Rigenos was the last living – and ageing now, with only a daughter, Iolinda, to carry on his line. Old and weary with hate – but still hating. Hating the unhuman folk whom he called the Hounds of Evil, mankind's age-old enemies, reckless and wild; linked, it was said, by a thin line of blood to the human race – an outcome of a union between an ancient Queen and the Evil One, Azmobaana. Hated by King Rigenos as soulless immortals, slaves of Azmobaana's machinations.
>
> (*The Eternal Champion*, p. 12)

As before, we have the reporting structure *whom he called*, again apparently justified by the metaphor. This time we also have the inserted clause *it was said,* which actually withholds Erekosë's agreement to everything in the clause, including the characterisation of Azmobaana as the Evil One. It would, however, be sufficient justification for the insertion if affirmation was being withheld for the historical accuracy of the statement. Once again, then, we have a refusal to affirm Humankind's evaluation disguised as a refusal to assert what cannot be possibly proved. Finally, one notes the subtle difference between *Hated by King Rigenos as soulless immortals* and *Soulless immortals hated by King Rigenos.* The latter asserts that they are soulless; the former again merely asserts that Rigenos thinks them so.

The adjectives also support the double reading: *unhuman* is the stroke of one letter away from being *inhuman,* a clear marker of negative evaluation. *Wild* is associated with *flowers* and *animals* equally, but its close association with *reckless* here provokes a negative reading. Examination of 709 instances of *reckless* in my corpus shows that it is overwhelmingly associated with negatives: its most common lexical collocates are *driving* and *death,* and it occurs regularly in negative pairs, such as *reckless and destructive, reckless and scandalous, reckless and illegal,* and *regular and greedy.* On the other hand, *audacity, courage, energy, heroism, honesty* and *idealistic* all also occur in the immediate environment of the word in my corpus. So once again, the text invites us to go along with Humanity's characterisation of the Eldren as bad while allowing itself loopholes through which it can escape on a second reading.

We have shown why it is that we go along with the original identification of Humanity as on the side of good – the problem–solution pattern is subtly evoked at the outset and lures us into a false understanding of the situation – and we have seen that the language in which the problem is first articulated is carefully constructed to ensure that false judgements are not averred. The question remains: why do we not regard Erekosë as the ultimate villain? Again, the clue lies in the wording of the initial paragraph of the Prologue quoted at the outset of this paper. I want to conclude by looking at a cluster of clauses from this paragraph. The first is:

I went to them

In my 100-million-word corpus there were just four instances of the pattern

x [human] *went to them*

and in all four cases the person going is in some respect dependent upon 'them' for help, support or custom. In other words, the people represented as 'them' have the upper hand. An example is:

> In the beginning it was Muslims driven out by Serbs who went to them for shelter.

Examination of the similar pattern 'x (human) *went to him*' finds four out of seven instances having exactly the same semantic implication: the person referred to as 'him' is the superior and the person going to him is going for help, support and so on, for example:

> My great-grandfather was the head of the clan and when there was a problem within the family, we went to him.

The same proves true of the pattern 'x (human) *went to her*'. Of five instances, four have 'her' as the source of succour or support.

So when Moorcock writes in his prologue:

> *They called for me and I went to them*

he is not only setting up a problem–solution pattern, he is using a structure associated with people of lesser power. By a kind of linguistic sleight of hand, he is telling us that the character who will bring about an apocalypse in the form of the extinction of humanity and the near-extinction of the Eldren is weaker than the callers. This weakness is indicated further in the following sentence,

> *I could not do otherwise*

which, of course, suggests that his actions were not of his choice, a special kind of weakness. There may, however, be another, more interesting implication to this sentence. There is only one example of this utterance in my corpus, but it is perhaps worth quoting in a fuller context than I have hitherto done with examples from my corpus:

> A founding member, he eventually became leader of the Italian Communist Party in his early thirties, fighting ludicrous factionalism as buttons were busily being sewn on so many black shirts. Given a 20-year sentence for his belief (from England, only this paper [The Guardian] was at his trial) he would continue to write

> to his mother, unaware that she had perished while he was inside. Gramsci died two days after his release. His father on his death bed, a couple of weeks later, would read the martyr's words over and over again: '... I could not do otherwise ... sons must sometimes cause great grief to their mothers if they wish to preserve their honour and dignity as men.'

No safe conclusions can be drawn from a single example, but I find it suggestive and in tune with my intuitions that the sentence *I could not do otherwise* in this example is associated with martyrdom and great heroism. It may therefore be that in three successive sentences Erekosë describes himself as hero/helper, as weak/powerless subject and as suffering martyr, all positions that he adopts in the course of the book.

The way I worded the previous point was deliberate: I could have said 'is portrayed' in place of 'describes himself', and, indeed, such a verb phrase would have been a more accurate representation of the fact, but I want to draw attention to the fact that Moorcock has articulated these apparently contradictory positions within Erekosë's voice. As noted at the beginning of this paper, Erekosë is seen as hero throughout the book, despite the fact that his actions are morally equivalent to those of the perpetrators of the Holocaust. Indeed in the Epilogue, Moorcock has Erekosë say:

> We cleansed this Earth of human kind. (*The Eternal Champion*, p. 158)

Moorcock's choice of word is uncomfortably prescient, given the recent horrors of ethnic cleansing in what used to be Yugoslavia.

I have argued that we accept him as hero initially because he is associated with the problem–solution pattern, and that we associate him with that pattern because of the choice of *called for* rather than *called to* or *called*. Now I want to argue that we accept his excuses later in the book because from the outset we accept his own characterisation of himself as not powerful and as suffering martyr. But in G. K. Chesterton's Father Brown story 'The Actor and the Alibi', from the collection *The Secret of Father Brown*, Father Brown draws attention to the fact that the general good opinion granted to one of the characters entirely stems from people's taking her own assessment of herself at face value. Perhaps Erekosë's words require a Father Brown as interpreter. In the Epilogue, he states explicitly the position hinted at in the prologue:

> I feel more certain than ever that it was not my decision. (*The Eternal Champion*, p. 158)

Most of the monsters of the twentieth century have made a similar excuse.

Notes

1. Michael Moorcock *The Eternal Champion* (London: Mayflower, 1970), p. 157. Subsequent page references in text.
2. Jacqueline L. Harvey 'A Sting in the Tale: An Examination of the Features of Surprise Ending Narratives' (Unpublished M.Phil. Thesis, University of Birmingham, 1996).
3. Mike Scott, WordSmith Tools (2nd edition, Oxford: Oxford University Press, 1997).
4. Genesis 21: 17, *The Holy Bible, New International Version* (1973).
5. The * indicates all the possible endings for the word, i.e. *-ing, -s, -ed,* and also no ending at all.
6. Vladimir Propp, *The Morphology of the Folktale* (1928; Austin: University of Texas Press, 1968).
7. The approach I am going to describe was first outlined by Eugene Winter, 'Fundamentals of information structure: a pilot manual for further development according to student need', mimeo, The Hatfield Polytechnic (1976) and developed in greater detail by colleagues and students of his. These include the present author in his *Signalling in Discourse*, Discourse Analysis Monographs No. 6, (Birmingham: ELR, The University of Birmingham, 1979); *On the Surface of Discourse* (London: George Allen & Unwin, 1983, reprinted by Department of English Studies, University of Nottingham, 1991); and 'Overlapping patterns of discourse organization and their implications for clause relational analysis in problem–solution texts', in C. Cooper and S. Greenbaum (eds.), *Studying Writing: Linguistic Approaches (Written Communication Annual 1)* (London: Sage, 1986); Michael Jordan in his paper 'Short texts to explain problem–solution structures and vice versa' in *Instructional Science* 9, 221–252, (1980), and his book *Rhetoric of Everyday English Texts* (London: George Allen & Unwin, 1984); and Winifred Crombie, in her books *Process and Relation in Discourse and Language Learning* and *Discourse and Language Learning: a Relational Approach to Syllabus Design*, both published by Oxford University Press in 1985.
8. Genesis 21: 15–20, *The Holy Bible, New International Version.*
9. For discussion of this possibility, see John McH. Sinclair, 'Planes of discourse', in S. N. A. Rizvi (ed.), *The Two-fold Voice: Essays in Honour of Ramesh Mohan* (Salzburg: University of Salzburg, 1981), pp. 70–89.

12

Storm, Whirlwind and Earthquake: Apocalypse and the African-American Novel

A. ROBERT LEE

I

> We need the storm, the whirlwind, and the earthquake ... What to an American slave, is your 4th of July? I answer: a day that reveals more to him, more than all the other days in the year, the gross injustice and cruelty to which he is the constant victim. To him, your celebration is a sham; your boasted liberty, an unholy licence; your national greatness, swelling vanity; your sounds of rejoicing are empty and heartless; your denunciations of tyrants, brass fronted impudence; your shouts of liberty and equality, hollow mockery; your prayers and hymns, your sermons and thanksgivings, with all your religious parade, and solemnity, are, to him, mere bombast, fraud, deception, impiety, and hypocrisy – a thin veil to cover up crimes which would disgrace a nation of savages. There is not a nation on the earth guilty of practices, more shocking and bloody, than are the people of the United States at this very hour.
>
> Frederick Douglass, 'What To The Slave is the Fourth of July?: An Address Delivered in Rochester, New York, on 5 July 1852'[1]

> Whilst we are bordering on a future of brighter things, we are also at our danger period, when we

must either accept the right philosophy, or go down
by following deceptive propaganda which has hemmed
us in for many centuries.

Marcus Garvey, 'The Future As I See It' (1923)[2]

The Black Artist's role in America is to aid in
the destruction of America as he knows it.

LeRoi Jones/Amiri Baraka, 'State/meant' (1965)[3]

On the faith of an eye-wink, pamphlets were stuffed
into trouser pockets. Pamphlets transported
in the coat linings of itinerant seamen, jackets
ringwormed with salt traded drunkenly to pursers
in the Carolinas, pamphlets ripped out, read aloud:
Men of colour, who are also of sense.
Outrage. Incredulity. Uproar in state legislatures.

We are the most wretched, degraded and abject set
of beings that ever lived since the world began.

Rita Dove, 'David Walker (1785–1830)' (1980)[4]

Frederick Douglass, whose oratory along with his landmark *Narrative* (1845) of slave life in and then escape from the plantations of Maryland made him one of the ascendant black voices of abolition, implies a Final Reckoning for the slaveholding America just having celebrated Independence Day 1852.[5] Marcus Garvey, 1920s Harlem luminary and prime mover in the 'Back To Africa' nationalist movement as embodied in his Universal Negro Improvement Association (UNIA), and then in the ill-fated Black Star Line by which he thought to take returnee blacks to Liberia and elsewhere in the 'mother continent', envisages worlds at binary racial opposites.

Imamu Amiri Baraka, his name newly Islamized from the LeRoi Jones of a Newark, New Jersey upbringing and recent Greenwich Village Beat phase, publishes 'State/meant' the same year, 1965, as his arrest by the FBI for allegedly using Federal funds intended for his Harlem theatre-work to build a gun arsenal in furtherance of Black Power. Rita Dove, African-American Poet Laureate (1993–95), looks back in her biographical poem of 1980 to the black Boston clothier-author of an 1820s revolutionary pamphlet literally smuggled in the pockets and linings of used garmentry into the slave South and beyond: *Walker's Appeal in Four Articles; Together with a Preamble, to the Coloured Citizens of the World, but in Particular and Very Expressly, to Those of the United States of America* (1829).[6]

All of these voices give witness and representative continuity to the seams of apocalypse within the literature of Afro-America. They also introduce a context for the black-written novel which, if likewise to be granted every variety of other interest, has long found itself drawn to a vision of worlds turned (or to be turned) upside down by slavery and its aftermath. The tradition runs from Martin R. Delany's inaugural story of would-be black Dixie and Cuban insurrection, *Blake; or the Huts of America; a Tale of the Mississippi Valley, the Southern United States, and Cuba,* issued in his lifetime only as instalments in the *Anglo-American Magazine* in 1859 and the *Weekly Anglo-African* in 1861–62, right through to, and beyond, Toni Morrison's slave-haunted, elegiac *Beloved* (1987).[7]

Whatever their differences of styling, the novels involved bespeak an American racial order brought to the edge of, if not actually seen to have crossed into, quite momentous unorder. Theirs has been the imagining of an Atlantic dispensation however much black alongside white, then also, and at the same time, black against white and vice versa. The black men and women whose stories they tell find themselves implicated not only in subterfuge, or escape, or exile, but often in the still larger shadow, and call, of apocalypse. Given the huge, dramatic oppositions built into, and perpetuated by, black-white American history, in one sense could it ever have been otherwise?

An enduring trope for this contradictory, massively spiralled drama duly appears in the first-ever published African-American novel (and by an author himself an ex-slave), William Wells Brown's *Clotel* (1853).[8] Having offered his account of Clotel as 'tragic mulatto', the daughter of Thomas Jefferson and his slave mistress driven to her own drowning in the Potomac, Brown calls attention to a begetting irony at the heart of the American story. In 1620 the freedom-seeking English-pilgrim *Mayflower* lands in Massachusetts. A year earlier, in 1619, a Dutch man-of-war sells 'twenty negars' to Captain John Smith in Jamestown, Virginia. However much the backward glance, and as Brown shrewdly emphasizes, these near-simultaneous events bear also a dark forward prophecy, ships and voyagers as the emblems of a paired and yet at the same time ominously unpaired America.

The footfalls which have followed become, in turn, as figural as literal, a kind of memorial, and often enough apocalyptic iconography of African-American people unfree in a land of American freedom and caught in each overlapping turn of 'the peculiar institution' and its continuity as 'race'. Nor is this to avoid recognition of the dangers in racializing all oppression or powerlessness. But from Jamestown in the 1600s to the Civil War in the 1860s, and from the 1920s as in equal part an era of Klan violence or city riot as Jazz Age, to the turbulent 1960s of Civil Rights and Black Power, in which Malcolm X would deliver his celebrated 'ballot or bullet' speech and Eldridge Cleaver his 'We shall have our manhood. We shall have it or the earth will be levelled by our attempts to gain it', a body of presiding terms enter and suffuse not only the novel, verse and drama (along with Afro-America's dazzling compendium of oral folklore and signifying), but, in truth, all American historical culture.[9]

All these eras carry markers of apocalypse as much personal as communal. From slavery-times the recurrence is to chains, shipholds, the dire variations of 'negro' into 'nigrah' or 'nigger', auction, Jim Crow, 'massa', lynch, miscegenation, cotton, the quarters, patrollers, John Brown, the Mississippi of 'sold down the river' and, inevitably, the Civil War, often referred to summarily as 'the war'. Each has become sedimented within black idiom, a lexicon of remembrance and accusation.

'Middle Passage' (1966), Robert Hayden's beautifully crafted poem-sequence, taps into this language as, in its own way, does Alex Haley's *Roots* (1976), and its massively watched TV offspins (*Roots*, 1977, and *Roots*: *The Next Generations*, 1979) – the New World as the black genealogy descended from Kunta Kinte and put to survive within and beyond an America of slave bondage.[10] But if an even earlier pathway into the writing of black apocalypse is to be sought it can be found in two respectively antebellum and postbellum African American novels.

II

> 'I now impart to you a great secret... I have laid a scheme, and matured a plan for a general insurrection of the slaves in every state, and successful overthrow of slavery!'[11]

So, in *Blake*, Martin Delany's Cuban-born hero, Henry Blake, gives notice to his fellow Mississippi slaves of his conspiracy to rid all the Americas of black bondage. It strikes an insurrectionary, and indeed an apocalyptic, note, a whole hemisphere called to account and, as need arises, brought into a war of human redress ('I am for war – war upon the whites,'[12] Blake subsequently announces to his Cuban followers of colour).

The novel, undoubtedly, has its *longueurs*. The plot can be maze-like, episodic, with lacklustre dips in style. Blake himself can smack of over-heroicization. The story, too, remains unfinished, a kind of promissory note. But at the same time there can be no doubting the implication of black epic, a war-to-be uncompromisingly given to the overthrow of tyranny. Further, it is conducted to wholly black auspices and without a John Brown at Herper's Ferry to aid and abet.

Each increment of the story implies tension, a gathering historical pace. Blake's immediate family is sold, arbitrarily, by the white plantation dynasty of Colonel Franks with a rescue narrative to follow. The creation of Blake's secret network across the slave states, be it Mississippi or Texas, Charleston or New Orleans, all with their respective slavery stories, ensures diversity along with width of canvas. The conspiracy of Cuban mulattoes and blacks first for their own disenslavement from the Spanish, and then, with a highjacked slaveship, to free slaves in the American south, gives off a sense of impending redemption at any or all cost. If a founding novel is to be invoked for black apocalypse it assuredly lies in *Blake*.

In Sutton Griggs's *Imperium in Imperio* (1899) the drama lies in separatism, an envisioned black-masonic takeover of Texas as the model 'to secure the freedom of the enslaved negroes the world over'.[13] Under the dual leadership of the militant Bernard Belgrave, a mulatto, and Belton Piedmont, a full black, who from principle reneges in favour of a gradualism he pays for with his death ('I am not for internecine war'),[14] the Imperium's founding is deliberately pitched to call into mind America's previous will to liberation. An address to the membership proclaims 'If it calls for a Valley Forge, let us be free.'[15] Like the novel itself, this serves as Griggs's warning speculation, his cautionary tale: continue the denial of black rights and America will risk a quite new War of Independence, an apocalyptic war of race.

III

That dream, or anti-dream, in fact begins early. Even so otherwise peaceable a founding slave poet as New England's Phillis Wheatley (1753–1784) contributes. Her 'Goliath of Gath' pietistically reworks the biblical epic of David and Goliath based on 1 Samuel XVII. Yet at the same time there can be no mistaking its more oblique pointer to a possible war of slave liberation ahead.[16] A pioneer text like *The Interesting Narrative of the Life of Olaudah Equiano, or Gustavus Vassa, the African* (1789) even more confirms the note, at once captivity and escape story, travelogue and proto-novel.[17] For as much as Equiano offers a life full of voyage and itinerary play of mind and belief, he at no time holds back on his own lived witness to slavery's incarceratory 'horror', as he calls it.

Slave narratives, estimated to run to a possible two thousand, yield a more explicit apocalyptic signature, be it as personal suffering, the religious dream of Zion, or, failing abolition and suffrage, the call for insurrection. Is not the Covey fight in Douglass's *Narrative* (1845), 'the turning point in my career', an apocalypse in small? How best to read the seven-year 'loophole' hiding out, separation from her children, and sexual menace of Dr. Flint as slaveholder, but as deepest trauma for the author-protagonist in Harriet Jacobs/Linda Brent's *Incidents in the Life of a Slave Girl* (1861)? Is not the well-known conciliatory temper of Booker T. Washington's *Up from Slavery* (1901) underwritten by a searing memory of bare childhood survival and indigency?[18]

Slave narrative, none the less, has not been the only vehicle of remembered insurrection. The *Amistad* shipboard mutiny of 1839 acts as a major source in Jones/Baraka's play *Slaveship. A Historical Pageant* (1967). By 1997 it had become a Steven Spielberg film. The insurrections of Denmark Vesey in South Carolina in 1822 and Nat Turner in Southampton County, Virginia, in 1831 caused massive panic, sometimes incredulity, throughout the slaveholding South. The latter, especially, supplies a continuing source of black spoken legend and, later, yet further controversy for its alleged sexual and other 'white' distortion when made over by William Styron into his 'meditation on history', *The Confessions of Nat Turner* (1967). Slavery into novel, moreover, looks back to quite the earliest antecedents. Frederick Douglass himself used the *Creole* affair of 1841, a slave-ship take-over led by Madison Washington, as the grounds for his adventuresome magazine novel *The Heroic Slave* (1853).[19]

It falls to Arna Bontemps, however, the 1920s 'New Negro' literary stalwart and Fisk librarian, to have written quite one of the most striking fictionalizations of black revolution. In *Drums at Dusk* (1939) he had chronicled Toussaint de l'Ouverture's leadership of Haiti's War of Independence in 1802 (to be given another canny intertextual reprise in Ralph Ellison's story, 'Mister Toussan', in 1941). But in *Black Thunder: Gabriel's Revolt: Virginia: 1800* (1936) he looks to a major uprising on American soil itself, Gabriel Prosser's brave, would-be apocalyptic slave seizure of Richmond, Virginia in 1800.[20]

'Hell's loose'[21] observes a white Jacobin sympathizer of the gathering revolt. It throws a due light on the 'win or lose all' (p. 69) rising hatched by Gabriel as much at his enslavement by word or image as by slaveholding fact – 'I'm tired of being a devilish slave' (p. 103). Doubly betrayed by a massive downpour of rain as by the retainer-slave, Old Ben, whose fear of God's wrath at this uprising against slavocracy causes him to reveal the plot and its leadership, Gabriel sees himself to the end as following in the steps of Toussaint.

Bontemps depicts him as a man of 'large design' (p. 198), 'dream' (p. 213), with physique to match, the lover of Juba, but above all 'the first for freedom of the blacks' (p. 222). Having failed to win slave freedom in Virginia (and in a striking echo of Babo in Herman Melville's 'Benito Cereno') Gabriel indicates the rope which eventually hangs him as his only bequeathed explanation or 'talk' (p. 223). Silence thus becomes another style of apocalypse, the willed refusal to answer to a dispensation which itself has hitherto stolen from him and his fellow slaves the right as much to speak as act.

IV

The Civil War equally, and inevitably, bears its own apocalyptic nuance – the national fissure of North and South, John Brown's killings at Harpers Ferry in 1859, each major battle and slaughter and, of most immediate relevance, the sacrifice of black troops fighting under Union colours. In this respect a poem like the 'The Colored Soldiers' by Paul Laurence Dunbar (known as 'the black Robbie Burns'), with its eulogy to brave deaths in the name of freedom, can be read in company with a novel like Frances Ellen Watkins Harper's *Iola Leroy* (1892) which, despite its main theme of Free Negro life in Philadelphia, at the same time celebrates black Civil War military heroism.[22] Did not, too, this returning black soldiery carry always the silhouette of possible further reprise, more race-driven combat, given the broken promises of Reconstruction?

Nor for a moment was the war and its abolitionist lead-up ever exclusively male-centred. Sojourner Truth and Harriet Tubman, vintage anti-slavers, call (again apocalyptically?) for a double liberation: that of black women both as work and sexual property. Margaret Walker's *Jubilee* (1936), a deliberate riposte to Margaret Mitchell's *Gone With The Wind* (1936), tells one version of that story in the figure of Vyvry as black wife-mother survivor faced with both the racial and personal chaos of slavery. Ernest Gaines's *The Autobiography of Miss Jane Pittman* (1971) offers another, a Louisiana plantation life and womanhood carried down through traumatic slave ownership, the Civil War, Reconstruction and segregation into a 1960s of Civil Rights.[23]

Alice Walker's versions of apocalypse, especially when measured from her 'womanist' credo and affiliation with Zora Neale Hurston, have been of long standing: whether a novel of black sorority told against the backdrop of Civil Rights with its Freedom Rides, Birmingham bombings in 1963, and marches like Selma in 1965, in *Meridian* (1976), or of the rise from silence and the abuse of rape and incest of a black child-into-woman like Celie in *The Color Purple* (1982), or the pre-Columbian, shamanistic vision of an Africa both liberative yet, in its passed-down practices of female circumcision, also repressive, as in *The Temple of My Familiar* (1989). Gayl Jones's *Corregidora* (1975), alongside, also casts the story of black women as pre-eminently one of sexual memory. The apocalyptic ravage of the female body begun under a patriarchal Brazilian slaver finds its eventual but painful redemption and voice in the blues of Ursa Corregidora as the novel's singer-heroine.[24]

The Great Migration, which between the 1890s and the 1920s took an estimated four million blacks north from rural Dixie into Harlem, Chicago and other cities, becomes for all the Jazz Age hooplah and flourish also a tableau of grim industrial divides by class and colour. Few more vivid depictions exist than William Attaway's *Blood On The Forge* (1941), a kind of blues narrative of the passage (with due scenes of betrayal and riot) of three brothers as they leave sharecropper Kentucky for the satanic steelmills of Monongahala Valley, Pennsylvania.[25]

V

It might have been only a matter of time before black apocalypse would take its comic-satiric turn. George Schuyler's *Black No More* (1931) supplies the script, a fiercely absurd picaresque as a rogue geneticist develops a skin treatment which turns black into white. The Menckenite flavour is unmistakeable as the ways of American racial caste reduce to degree-zero absurdity. The world in which Mark Disher rises to control the colour-altering 'Black-no-more, Incorporated' turns, in a boldly mischievous echo of freedom's cry, on the phrase 'White at Last!', the supposed core longing of all (or at least all middle-class) Afro-America.

The results, indeed, veer into zaniest apocalypse: a mock Klan and NAACP caught up in unholy alliance; 'chromatic democracy' as the ruination of the black prosperity paradoxically won from segregated real estate or 'Afro' haircare (as personified in Madame Sisseretta Blandish); W. E. B. DuBois, James Weldon Johnson, Marcus Garvey and others of the 1920s 'New Negro' elite all put to satiric account; mulatto children born to formerly black but now 'white' women; ex-blacks several times whiter than 'authentic' whites, who so reverse into a new underclass; black confidence-men 'whites' lynched in an always lynch-ready Mississippi; and 'stained' skin as eventual chic modishness. If occasionally over-wrought *Black No More* yields an exquisite pillorying, it is the transformation of 'race' from tragic into comic farce.

VI

With citification, and allowing for the kind of lyric 'down-home' black metropolis depicted in, say, Claude McKay's *Home To Harlem* (1928), the vocabulary changes once again.[26] This time the register becomes one of ghetto, tenement, project, hustle, gang, shootout, penitentiary, not to mention openly spoken codes of verbal offence and defence for which a brute memory term like 'motherfucker' acts as a kind of emissary shorthand. And if any one novel gives focus to the city as *huis clos*, a tenement Chicago hell of 'fear, flight and fate' in its own three-section partition, it lies in Richard Wright's landmark *Native Son* (1941).[27]

Bigger Thomas's violence, from the opening episode with the rat through to the murder and incineration of Mary, Wright tacks to a whole traumatic landscape deep within. The rat's belly 'pulsed with fear', its 'black beady eyes glittering', anticipate Bigger himself made rodent, and then murderous, by the urban maze which regulates and dehumanizes him. Each of Wright's images for Bigger's life suggests a tension of enclosure and revolt, the family tenement, the snarling furnace in which he burns Mary's body, his hideouts from the police, and, finally, the prison-cell from which he goes to his execution with 'the faint, wry bitter smile' upon his face. Is this not a black apocalypse of ghettoed psyche as much as black city?

The narrative vein which picks up on Wright's vision (it has always been an inadequacy to think him merely a naturalist) includes, symptomatically, Ann Petry's *The Street* (1946) with its story of Lutie Johnson driven to murder in tenement Harlem, Jones/Baraka's *The System of Dante's Hell* (1965), a story-cycle novel of racist America as Inferno, and Gloria Naylor's *The Women of Brewster Place* (1982) with its case portraits of brute violence against women and its counter-dream of black female bonding.[28]

There can, too, be no overlooking the city as a related apocalypse of drug culture or 'smack' – heroin, cocaine and, lately, methamphetamine. Herbert A. Simmons's *Corner Boy* (1957) early inscribes an unsparing ghetto and drugs life in St. Louis, Missouri. George Cain's *Blueschild Baby* (1967) makes the needle a virtual saviour-devil instrument, with the sway of 'horse' (or heroin) as relief yet always a further dependency. Hal Bennett's *Lord of Dark Places* (1971) links drug habit into black existential pain, its will to catharsis drawn into the novel's graphic, emeticizing prose. Robert Deane Pharr's *S.R.O.* (1971) turns upon 'Single Room Occupancy' Harlem as the centre of a downward spiral of addiction presided over by the grim, deathly supplier Sinman. Latterly Ray Shell's *Iced* (1993), told in italicized line-measures, unfolds a regime of crack cocaine virtually Dostoevskian in its willed, violent abandonment of anything but the dictates of drug habit.[29]

The city under riot, relatedly, becomes a literary continuity in its own right. Charles Chesnutt's *The Marrow of Tradition* (1901), based on the uprisings against Jim Crow in Wilmington, North Carolina in 1898, offers a prologue. Walter White's *Flight* (1926), though better known as a novel of 'passing', draws upon white Atlanta's abuse of its black citizens in the 1920s. Ralph Ellison's *Invisible Man* (1952), among all its other workings of myth and blues, transforms the 1940s Harlem riots into surrealized pitched battle between The Brotherhood and the followers of the Garveyite 'Ras The Destroyer; the upshot is a startling, and again memorial, vision of Afro-America's house driven into division against itself.[30]

In *Sons of Darkness, Sons of Light: A Novel of Some Probability* (1969) New York becomes the scenario for a narrative of black-white armed confrontation, a latterday near-armageddon in the 1960s of clenched-fist black activism across Brooklyn, Manhattan and Harlem. That vision reads with still greater subtlety in Williams's *The Man Who Cried I Am* (1967), his adroitly modernist political thriller spanning America, Europe and Africa in which a Richard-Wright-like exile is killed under the genocidal 'King Alfred Plan' of a racist international cadre known as the *Alliance Blanc.* In Leon Forrest's *The Bloodworth Orphans* (1977) America is imagined as hemispheric racial orphanage, and in language to recall the narrative high baroque of Jean Toomer's *Cane* (1923), a world hexed into citied race-war rather than healed by its multiple cross-ties of blood and clan.[31]

No account of urban apocalypse in the African-American novel would be complete, however, without invoking Chester Himes and, above all, his Harlem 'domestic tales' or *romans policiers.* As his case-hardened detective duo, Coffin Ed Johnson and Grave Digger Jones, take on each about-face and often quite irreal crime from *For Love of Imabelle* (1957) through to *Blind Man With a Pistol* (1969) and the posthumous *Plan B* (1993) – in which the one cop kills the other – so it becomes clear that Himes is about a far greater unravelling. How did so massive, contradictory, and dangerous a *comédie humaine* as Harlem ever came into being? That each of his stories draws upon a wickedly funny gallows humour, along with plotlines which edge towards apocalyptic maze, makes for its own bonus. For in Himes's Harlem world of hustlers and con-men, chippies and religious freaks, 'race' has veered into 'black' black comedy, a reeling, absurdist city of nightmare.[32]

VII

Two further novels, each its own *tour de force,* offer a closing perspective. John A. Williams's *Captain Blackman* (1972) works as a kind of serial history in which Captain Abraham Blackman, under frontline fire in Vietnam, fantasizes black warriordom from the American Revolution through to the Nuclear Age. From a figure like the black Crispus Attucks, the first to be killed in the war against the British, Blackman passes on to a succession of black soldering through the War of 1812, the Civil War, the Plains War, the Spanish American War, World Wars I and II, Korea and Vietnam, with an Epilogue in which Blackman's Aryan antagonist, General Ishmael Whittman, is out-finagled by a black takeover of America's entire nuclear facility. History, finally, so falls under alternative racial auspices, with all relevant inference left to the reader.[33]

Toni Morrison's *Beloved* (1987) re-imagines the apocalypse of slavery as a haunting, Sethe Suggs's killing of her own daughter to save her from a life of slave bondage in antebellum Kentucky. 'Beloved' returns to take over Sethe and to possess the family's Ohio home as retribution, a vision of death over life eased only by Sethe's love for Paul B, another returnee slave. The novel bespeaks trauma and horror, but also love, the yet-one-more-memorial apocalypse of a slave mother driven to destroy her own progeny (and so, in large measure, herself) in the face of slavery's own destructive writ.[34]

The term 'apocalypse', to be sure, requires its latitude. Even the most pertinent novels do more than offer all-or-nothing doomsday scripts, whether the subject be slavery or uprising, a 1960s Watts or the videoed all-white LAPD squad beating of Rodney King and its fiery aftermath in Los Angeles's South Central in April 1992. None the less, one black fictional landscape of apocalypse does come close, that of Science Fiction, and for which not only George Schuyler's *Black No More* but still more his *Black Empire* (1938), written under the name Samuel I. Brooks and a global plague novel in which a colonized Africa battles a colonizing Europe, points the way.

Among contemporaries, Ishmael Reed, ranking black wit and metafictionist, can so be enlisted on the basis of his 'space satire' of Nixonian America and its racism in *The Free-Lance Pallbearers* (1967), not to mention his treatment of slaveholding as postmodern fantasy in *Flight To Canada* (1976). In Afro-America's more recognizable SF or 'speculative' writing, a sense of apocalypse in gender as much as race comes into play, be it Samuel R. Delany's four-volume saga of master–slave sexuality and power fetish begun in his *Tales of Neveryon* (1979) or Octavia E. Butler's 'slave-plantation' female survival parables in novels like *Patternmaster* (1976) and *Mind of My Mind* (1977).[35]

Whether 'black apocalypse' is best thought of as a genre of novel in its own right, or more a sub-genre, one underlying issue can be seen to persist. From *Blake* to *Captain Blackman, Imperium in Imperio* to *Beloved,* this has been a fiction as taken up with Afro-America's remembrance of its future as of its past or present, apocalypse then, or now, but only, and equally, as if always still to come.

Notes

1. Reprinted in Henry Louis Gates Jr. and Nellie Y. McKay (eds), *The Norton Anthology of African American Literature* (New York and London: W. W. Norton & Company, 1997), p. 388.
2. Reprinted in Amy Jacques-Garvey (ed.), *The Philosophy and Opinions of Marcus Garvey* (New York: Atheneum, 1969).
3. LeRoi Jones/Imamu Amiri Baraka, 'State/meant' in *Home: Social Essays* (New York: William Morrow and Company, 1966).
4. Rita Dove, 'David Walker (1785–1830)', in *Selected Poems* (New York: Pantheon/Vintage, 1993).
5. Frederick Douglass, *Narrative of the life of Frederick Douglass, an American slave. Written by himself* (Boston: Anti-Slavery Office, 1845). Reprinted in Henry Louis Gates, Jr. (ed.), *The Classic Slave Narratives* (New York: Signet/Penguin, 1987).
6. David Walker, *Walker's Appeal in Four Articles; Together with a Preamble, to the Coloured Citizens of the World, but in Particular and Very Expressly for Those of the United States* (Boston: David Walker, 1829). The standard edition is Charles M. Wiltser (ed.), *Walker's Appeal* (New York: Hill & Wang, 1965).
7. For the first reconstituted edition of Delany's novel, see Floyd J. Miller (ed.), *Blake or The Huts of America* (Boston: Beacon Press, 1970); Toni Morrison, *Beloved* (New York: Alfred A. Knopf, 1987).
8. William Wells Brown, *Clotel or, The President's Daughter, A Narrative of Slave Life in the United States* (London: Partridge & Oakey, 1853). The novel was twice revised: *Clotelle: A Tale of the Southern States* (Boston: James Redpath, 1964); *Clotelle: Or, The Colored Heroine. A Tale of The Southern States* (Boston: Lee and Shepherd, 1867). Reprinted in William L. Andrews (ed.), *Three Classic African-American Novels* (Mentor/ Penguin, 1990).
9. Malcolm X, 'The Ballot or Bullet', transcribed and republished in Henry Louis Gates Jr. and Nellie R. McKay (eds), *The Norton Anthology of African American Literature* (New York and London: W. W. Norton & Company, 1997); Eldridge Cleaver, *Soul on Ice* (New York: McGraw Hill, 1968).
10. Alex Haley, *Roots* (New York: Doubleday, 1975).
11. Martin R. Delany, *Blake or The Huts of America* (reprinted Boston: Beacon Press, 1970), p. 39.
12. *Blake* (Beacon Press), p. 290.
13. Sutton Griggs, *Imperium in Imperio* (Cincinnati, Ohio: Editor Publishing Company, 1899). Republished as *Imperium in Imperio* (New York: Arno/The New York Times, 1969), p. 191.
14. Ibid., (Arno), p. 242.
15. Ibid., (Arno), p. 221.
16. Phillis Wheatley, *The Collected Works of Phillis Wheatley*, ed. by John Shields, *Schomburg Library of Nineteenth-Century Black Women Writers* (New York: Oxford University Press, 1988).
17. Olaudah Equiano, *The Interesting Narrative of the Life of Olaudah Equiano, or Gustavus Vassa, the African, Written by Himself* (London: Printed for

and Sold by the Author, No. 10, Union-Street, Middlesex Hospital; and may be had of all the Booksellers in Town and Country. Entered at Stationers' Hall, 1789); *The Interesting Narrative of the Life of Olaudah Equiano or Gustavus Vassa, the African* (London: 1790); *The interesting narrative of the life of Olaudah Equiano, or Gustavus Vassa, the African, written by himself* (New York: Printed and Sold by W. Dureell at his book-store and printing office, No. 19, Q Street, 1791). 2 vols in 1. For a current edition see Henry Louis Gates Jr. (Ed.), *The Classic Slave Narratives* (New York: Penguin 1987).

18. Frederick Douglass, *Narrative* (Boston: Anti-Slavery Office, 1845); Harriet Jacobs/Linda Brent, *Incidents in the life of a slave girl, written by herself*. Edited by L. Maria Child (Boston: Published by the author, 1861); Booker T. Washington, *Up From Slavery, An Autobiography* (New York: A. L. Burt Co., 1901).
19. LeRoi Jones/Imamu Amiri Baraka, *Slaveship. A Historical Pageant* (1967) in *Four Black Revolutionary Plays* (Indianapolis: the Bobbs-Merill Co., 1969); William Styron, *The Confessions of Nat Turner* (New York: Random House, 1967) – for the controversy the novel aroused see John Henrik Clarke (ed.), *William Styron's Nat Turner: Ten Black Writers Respond* (Boston: Beacon Press, 1986); and Frederick Douglass, *The Heroic Slave*, originally published in March 1853 in *Frederick Douglass's Newspaper*, and then as *The heroic slave; a thrilling narrative of the adventures of Madison Washington, in pursuit of liberty* (n.p. 1863). The text is reprinted in William A. Andrews (ed.), *Three Classic African-American Novels* (New York: Penguin/Mentor, 1990).
20. Arna Bontemps, *Drums at Dusk* (New York: Macmillan, 1939) and *Black Thunder: Gabriel's Revolt; Virginia 1800* (New York: Macmillan, 1939).
21. *Black Thunder: Gabriel's Revolt: Virginia: 1800*, reprinted as Beacon Paperback No. 305 (Boston: Beacon Press, 1968), p. 130. All page references are to this edition.
22. Paul Laurence Dunbar, *The Complete Poems of Paul Laurence Dunbar* (New York: Dodd, Mead, 1913); Frances Ellen Watkins, *Iola Leroy, or Shadows Uplisted* (Philadelphia: Garrigues, 1893).
23. Margaret Walker, *Jubilee* (Boston: Houghton, Mifflin, 1966); Ernest Gaines, *The Autobiography of Miss Jane Pittman* (New York: Dial Press, 1971).
24. Alice Walker, *Meridian* (New York and London: Harcourt, Brace, Jovanovitch, 1976), *The Color Purple* (New York and London: Harcourt, Brace, Jovanovitch, 1982), *The Temple of My Familiar* (New York and London: Harcourt, Brace, Jovanovitch, 1989); Gayl Jones, *Corregidora* (New York: Random House, 1975).
25. William Attaway, *Blood on the Forge* (New York: Doubleday, Doran, 1941).
26. Claude McKay, *Home To Harlem* (New York and London: Harper, 1928).
27. Richard Wright, *Native Son* (New York: Harper, 1940).
28. Ann Petry, *The Street* (Boston: Houghton Mifflin, 1946); LeRoi Jones, *The System of Dante's Hell* (New York: Grove Press, 1965); and Gloria Naylor, *The Women of Brewster Place* (New York: Viking Press, 1982).

29. Herbert Simmons, *Corner Boy* (Boston: Houghton Mifflin, 1957); George Cain, *Blueschild Baby* (New York: Doubleday, 1970); Hal Bennett, *Lord of Dark Places* (New York: Bantam, 1971); Robert Deane Pharr, *S.R.O.* (New York: Doubleday, 1971); and Ray Shell, *Iced* (New York and London: Harper Collins, Flamingo, 1993).
30. Charles Chesnutt, *The Marrow of Tradition* (Boston and New York: Houghton, Mifflin, 1901); Walter White, *Flight* (New York: A. A. Knopf, 1926); Ralph Ellison, *Invisible Man* (New York: Random House, 1952).
31. John A. Williams, *Sons of Darkness, Sons of Light: A Novel of Some Probability* (Boston: Little, Brown, 1969) and *The Man Who Cried I Am* (Boston: Little, Brown, 1967); Leon Forrest, *The Bloodworth Orphans* (New York: Random House, 1977); Jean Toomer, *Cane* (New York: Boni and Liveright, 1923).
32. Chester Himes, *For Love of Imabelle* (Greenwich, Connecticut: Fawcett, 1977); *Blind Man With A Pistol* (New York: William Morrow, 1969); and *Plan B* (Jackson, Mississippi: University Press of Mississippi, 1993).
33. John A. Williams, *Captain Blackman* (Garden City, New York: Doubleday, 1972). Other key novels in this tradition include William Gardner Smith, *The Last of the Conquerors* (New York: Farrar, Straus, 1948); John O. Killens, *And Then We Heard The Thunder* (New York: Knopf, 1963); and Sam Greenlee, *The Spook Who Sat By the Door* (New York: Bantam, 1969). A most useful historical account is to be found in Nat Brandt, *Harlem at War: The Black Experience in WWII* (Syracuse, New York: Syracuse University Press, 1996).
34. Toni Morrison, *Beloved* (New York: Alfred A. Knopf, 1987).
35. Ishmael Reed, *The Free-Lance Pallbearers* (Garden City, New York: Doubleday, 1967) and *Flight to Canada* (New York: Random House, 1976); Samuel R. Delany, *Tales of Neveryon* (New York: Bantam, 1979); Octavia E. Butler, *Patternmaster* (New York: Avon, 1976) and *Mind of My Mind* (New York: Avon, 1977).

13

Stylish Apocalypse: Storm Constantine's *Wraeththu* Trilogy

VAL GOUGH

The opportunity afforded by science fiction and science fantasy to envisage post-apocalyptic cultures is an attractive one for any writer interested in imagining post-patriarchy. Storm Constantine's *Wraeththu* trilogy (1987–8) is a notable example of the way that the fictional depiction of 'mankind's funeral' (I, pp. 140–1)[1] can be used as the means to challenge a range of dominant cultural assumptions to do with gender, sexuality and subcultural values. Constantine replaces the familiar motif of sudden ecological or nuclear disaster with a less quantifiable but no less devastating apocalypse: 'Not the final sudden death we all envisaged, but a slow sinking to nothing' (I, p. 10). Like other female science fiction and science fantasy writers concerned with the implications of patriarchy, Constantine sees the causes of the 'funeral' residing in man's own gender dominance: 'Man burned himself out from within. He had no balance; without it he perished' (III, p. 247).[2] Across volumes entitled *The Enchantments of Flesh and Spirit* (1987), *The Bewitchments of Love and Hate* (1988), and *The Fulfilments of Fate and Desire* (1988), she depicts the extinction of men – and the survival of women – in a twenty-first-century world increasingly dominated by hermaphrodite mutants called Wraeththu.[3] No longer human, Constantine's Wraeththu (also known as Hars) have developed a sophisticated caste system based on spiritual and psychic powers, taught largely through sexual training or 'aruna'. Biologically they combine male and female genitalia, yet having mutated from men, they continue to favour masculine aspects of behaviour, and this provides Constantine with much opportunity for debate about gender and sexuality. But Constantine is equally fascinated by the subversive potentials of British subcultures and their cultural and aesthetic styles. Thus images familiar from numerous other apocalyptic science fiction texts – abandoned, rusting cars; strewn corpses; derelict cities – contrast starkly with the subcultural stylishness of burgeoning Wraeththu culture. The trilogy combines largely familiar fantasy and science fiction elements, but what makes the series so compelling – and what I shall focus on in this discussion – is precisely this close attention to *style*, that eminently 1980s concern both in popular culture and in literary theory. The apocalyptic 'coming' of the Wraeththu is the coming of style, and for Constantine, the meaning of apocalypse (from the Greek word *apocalypsis* meaning to uncover, reveal) is precisely the revelation of the subversive and redemptive potentials of subcultural style.

The invariably opulent, sensual and highly aestheticised lifestyles of the numerous Wraeththu tribes encountered across the trilogy provide extensive scope for compelling narrative description, from hairstyles to clothes to home decor. The quest motif generates a nomadic Wraeththu anti-hero, Cal, but more importantly, an opportunity for a rich travelogue of styles and customs, largely in the vein of the following:

> The Kakkahaar are steeped in mysticism [...] I expected them to lead an austere life, but in fact found them to be a luxury-loving tribe. They loved to be waited on, hungered for comfort and trinkets; their Ara servants were dressed in diaphanous silks and heavily hung with gold adornments [...] Lianvis, asking us polite questions about ourselves, but not too prying, led us to a tasselled pavilion; his home. Inside, it reminded me of Seel's living-room, though Seel would have been sick with envy had he seen it. The colour scheme was dark bronze, dark gold and black. Tall, decorated urns spouted fountains of peacock feathers, canopies hung down from a central pole sparkling with sequins. The tent was so large it had several different rooms. A near-naked Har with hair to his thighs bound with black pearls, rose from the couch. (I, p. 90)

Most of the Wraeththu were originally gay men, and it seems that Constantine is happy to draw on the familiar if somewhat clichéd image of gay male style-consciousness. But, mindful of Jacques Lacan's contemporary spin upon the well-known dictum 'style maketh the man' (implying the constitutive role of style in the formation of subjectivity), I want to argue that Constantine's rich fictional evocation of post-apocalyptic cultural and linguistic styles functions as an oppositional discourse within the context of a narrative where humanity's demise goes hand in hand with a lack of style. Thus newly mutated Pell remembers his human life as one of aesthetic deprivation and a lack of style which is made concomitant with the degeneration of the race:

> At the age of fifteen, I lived in a dusty, scorched town at the edge of a desert. I was the son of a peasant, whose family for centuries had worked the cable crop for the Richards family. Our town was really just a farm, and to call it that lends it an undeserved glamour... We lived in cruel, bitter, petty country and it was inevitable that we shared many of these characteristics. Only when I escaped did I learn to dislike it. Then, I existed in a mindless, innocent way, ignorant of the world outside our narrow territories and content to stretch and pound the cable fibre with the rest of my kind. I don't suppose I ever did really think about things. The closest I came to this was a dim appreciation of the setting sun dyeing all the world purple and rose, lending the land an ephemeral beauty. Even the eye of the true artist would have had difficulty in finding beauty in that place, but the sunsets were pleasantly deceptive. (I, pp. 9–10).

We are told later that the 'women are drying up' (I, p. 93) but aesthetic barrenness is presented as just as much a cause of humanity's end as reproductive sterility. On closer analysis, we shall find that Constantine's commitment to an aesthetic sensibility is grounded upon a politicised sense of the subversive, resistant functions of subcultural styles. I shall examine the particular subcultural styles Constantine draws upon in her vision of Wraeththu culture, en route to asking the question, how ultimately subversive of mainstream 'straight' assumptions is that very *stylish* apocalyptic vision?

Constantine chooses to locate the origins of her fictional superbeings in a disaffected male subculture: 'They said it was a youth cult, and then more than that' (I, p. 17), and it has long been recognised that visual styles (clothes, hairstyles, accessories etc.) are key elements in young people's expression, exploration and constitution of their own individual and collective identities.[4] Hence it has been argued that the style of subcultural groups functions as a 'grounded aesthetics' full of meaning and implication.[5] Indeed, in his influential book *Subculture: The Meaning of Style* (1988), Dick Hebdige argues that the emergence of youth subcultures in the post-war period reflects a breakdown of consensus and a challenge to cultural hegemony which is not expressed directly, but rather obliquely through style.[6] Thus the construction of subcultural style is 'a gesture of defiance or contempt', it 'signals a Refusal [*sic*]'.[7] Subcultural styles of dress, language etc. are thus always 'pregnant with significance', and they function as '"maps of meaning" which obscurely represent the very contradictions they are designed to resolve or conceal'.[8] Such an understanding of subcultural style implies that Constantine's close attention to the depiction of Wraeththu style is similarly pregnant with significance, revealing bodies, clothes, hairstyles, makeup, jewellery and language as profound signifiers of cultural meaning. Umberto Eco said 'I speak through my clothes',[9] and Constantine presents us with a race who are particularly obsessed with their attire:

> As with all the other tribes I had visited, he had brought me clothes. It is something that is almost a fetish with the Wraeththu. Wherever you go your clothes are replaced with the prevailing fashion. (I, p. 205)

Moreover, Constantine's implicit cultural referents in the trilogy are late-twentieth-century popular phenomena – punk, Gothic rock, David Bowie, the New Romantics movement and gay male culture – which all place particular emphasis on the effects of style. For example, as Simon Frith says:

> ... as a performer Bowie celebrated the cleverness of his own false positions, and his 1970s switches of style, from glam-star to the disco sounds of *Young Americans* and *Station to Station* to European studio art, were bound by a commitment to stylishness. The settings changed (from stadium to disco to club) but not the sense that every move that Bowie made depended on a purely aesthetic judgement.[10]

In her portrayal of Cal's spiritual journey (and his tortuous moves towards a reconciliation with his beloved, Pell) as a series of encounters with different styles of living, Constantine clearly shares this subcultural sense of 'the dramaturgy of life as a series of lifestyles'.[11]

The *Wraeththu* trilogy has been described as 'informed by a late punk/Goth sensibility',[12] yet this description hardly does justice to the range of diverse British subcultural styles which Constantine plunders, and the detailed way in which she does so. An anthropological interest in the tribal patterns forming after holocaust is a common theme of apocalyptic fiction, and a number of apocalyptic novels have depicted subcultures as forms of survival – for example, Doris Lessing's *Memoirs of a Survivor* (1974) and Angela Carter's *Heroes and Villains* (1969) – but rarely with such close attention to subcultural style. New Age elements characterise the first Wraeththu settlement we encounter, Saltrock, with its makeshift structures and lively ambience:

> ... the buildings were constructed of a mad variety of materials, with seemingly little organisation. Some were quite large and made of solid wood, others little more than thrown-together metal sheeting or mere tents and animal hides [...] The inhabitants, creatures as startling as Seel, exuded spirit and energy. (I, pp. 26–7)

Seel's chosen style mixes aristocratic and tribal elements in a typically subcultural manner:

> He who rode the thin horse skidded it to a halt in front of us. Pebbles flew everywhere. When he left from the animal's back, it was in a wild tangle of flying rags, tassels, and flying red, yellow and black hair. (Another reality shocked me cold as the sexes mingled. Was this creature male or female, or could it be both ... ?!). (I, pp. 25–6)

Seel's particular form of stylistic eclecticism and genderblending evokes the aristocratic tribalist style of British ex-punk singer Adam Ant, who, says Dave Laing,

> [went] from being a 'bad boy' of punk ... [to] the 'principal boy' of pop, retaining his penchant for dressing up in the process. Hardly changing his costume, he moved from playing the Marquis de Sade to Prince Charming. For a while, too, he even had his own subculture (the 'ant people'), in a pre-teen parody of punk itself.[13]

All Wraeththu are indeed 'principal boys' in their camp blending of gendered styles. Sleek muscular bodies combine with long, often elaborately styled hair and a penchant for luxurious clothes and ornate jewellery. Their 'shifting male/female ambience' (I, p. 29) owes much to punk androgyny – which was 'slim, slight and invariably arty'.[14] Indeed, the very word 'punk' has a history of usage as a term for deviant male sexuality, especially in the United States.[15] Many other subcultural styles of the 70s and 80s flirted with androgyny. The Bowie-boys of the mid-70s used stylised makeup to create sexual shock and ambiguity, the New Romantics projected a foppish form of masculinity, and Goths of both sexes espouse a morbid, vampiric, androgynous style.

The language of youth subcultures is evoked in Wraeththu speech, which is a pungent combination of high and low linguistic varieties, mixing contemporary youth slang with a ritualistic discourse to signal the subversion of usual class distinctions. The use of taboo words and slang evokes subcultural language styles, especially punk. Inclusion of fictional Wraeththu words not only reinforces the Wraeththu's status as mutants and hence quasi-aliens, but also gives the impression of their language as a form of subcultural argot, a secret code understandable only to the initiated. Thus the pre-inception Pell encounters Wraeththu's linguistic strangeness:

> 'He looked like... like, I don't know. He was a bit like Cal, only as dark as Cal is fair. High cheekbones, sulky eyes. In a way you remind me of him; the same temperament I think. That's probably why Cal is kelos over you. He and Zack were chesna.' 'Flick,' I said, shaking my head at him. 'What the hell are you talking about. You must know I don't understand half of it.' (I, p. 40)

The music of youth subcultures, particularly punk rock, provides the inspiration for Constantine's depiction of the 'Temple Radiant', otherwise a largely un-mutated night club:

> Primal and thrilling, the music roared through my head. Rue leapt around the other musicians; His voice was a scream then a snarl; he crouched to tease the nearest of his audience, leaping up; his body as supple as a snake. (I, p. 231)

Such subcultural elements are, however, located within a narrative context that to some extent romanticises and sanitises their original chaotic and rebellious aspects. Constantine melds subcultural street styles with a fantasy element of quasi-medievalised luxury and opulence, and this means that the shock effect associated with subcultures, particularly punk, is somewhat diminished. Hebidge sees subcultures as functioning as a challenge to cultural hegemony, but there is an element of escapism in Constantine's vision which involves collusion with some hegemonic values. Regression to a Dark Age or quasi-medieval or tribal mentality is a familiar motif of apocalyptic and post-holocaust novels – a notable example is Russell Hoban's *Riddley Walker* (1980) – but Constantine deliberately shields the reader from the unpleasant details of humanity's demise, which we tend to learn about second-hand. Instead, Constantine focuses her narrative largely on the ruling elite of the Wraeththu tribes, in all their power, luxury and privilege, and by doing so she slides away from addressing some of the issues most pertinent to disaffected youth: unemployment, poverty and abortion, for example. However, this is not to say that Constantine's use of subcultural styles is simply mere ornament or compelling atmosphere. Indeed, as we shall see, her trilogy *does* represent subcultural resistance as a viable form of social change, and she uses subcultural style as the means to address gender issues, aspects of popular creativity, and questions of gay sexuality.

By drawing on existing British youth subcultures in her creation of a fictional world, Constantine alludes to their often characteristically apocalyptic sensibility. Dick Hebdige has vividly described the peculiarly 'apocalyptic' summer of 1976 – a summer marked by a vicious heatwave and resulting drought – during which punk made its sensational debut in the music press:

> The heatwave was officially declared a drought in August, water was rationed, crops were failing, and Hyde Park's grass burned into a delicate shade of raw sienna. The end was at hand and Last Days imagery began to figure once more in the press.[16]

Thus, says Hebdige:

> It seems entirely appropriate that punk's 'unnatural' synthesis should have hit the London streets during that bizarre summer. Apocalypse was in the air and the rhetoric of punk was drenched in apocalypse: in the stock imagery of crisis and sudden change.[17]

Goth style is redolent with eschatological implications,[18] and the Bowie cult identified with music which has been described as 'simultaneously personal, apocalyptic and radical'.[19] The 'profane and terminal aesthetic' of Bowie and the New York punk bands[20] is reflected in the first volume of the trilogy where humans fear Wraeththu as profane and abject agents of apocalypse:

> It was said that it had started as small groups of youths. *Something* had happened to them ... Oh yes, they changed. They became something like the werewolves my grandmother remembered tales of. Spurning the society that had bred them, rebelling totally, haunting the towns with their gaunt and drug-poisoned bodies; all night-time streets became places of fear. They dressed in strange uniforms to signify their groups, spitting obscenities upon the sacred cows of men, living rough in all the shunned places ... (I, pp. 10–11)

Indeed, the Wraeththu look upon themselves as a 'cleansing fire' from which 'new things would grow' (I, p. 19). The Wraeththu's stylish 'coming' evokes the Book of Revelation and the resurrection of the dead on Judgement Day, for their practice of inception – the process by which men become Hars – is a kind of quasi-death followed after three days by rebirth. Indeed, Wraeththu think of themselves as perfected beings, who rise from the 'first death' of inception to a quasi-immortality:

> They had the power to change the sons of men to be like themselves. As with the first, within three days of being infected with Wraeththu blood, the convert's body has completed the necessary changes. Many of them develop extra-sensory faculties. All are a supreme manifestation of the combined feminine and masculine spiritual constituents present in Mankind. Humanity has abused and abandoned its natural strengths: in Wraeththu it begins to bloom [...] Hara are ageless. Their allotted lifespan has not yet been assessed, but their bodies are immune to cellular destruction through time. As they are physically perfect, so they must strive towards spiritual perfection. (I, p. 50)

Reinforcing the motif of resurrection is the figure of Pell, who is literally killed by the supernaturally powerful Thiede and reborn via the 'holocaust' of aruna (I, p. 202), as a supremely perfected being. Thus Constantine is concerned to fictionally explore the post-apocalyptic potentials of subcultures and their practices, and whereas in reality subcultures are ultimately assimilated back into mainstream culture – through media trivialisation and mass commodification – fiction enables Constantine to imagine an alternative scenario, where the subversive subculture brings about the *end* of the dominant culture.

But despite the fact that Wraeththu sometimes appear 'like something from the Apocalypse', like 'messengers of Death's angel' (I, p. 159), Constantine is careful to keep ambiguous their actual role in the decline of mankind. The true causes of men's gradual extinction are never pinpointed. Moreover, the 'coming' of the Wraeththu is not portrayed as unproblematic and, indeed, Wraeththu themselves are depicted as questioning the value of their subcultural resistance:

> What are we? How? Why? To what end? It is more than just a fun time, running wild and screaming, 'Hey, let's get the bastards that fucked the world up!' It has to be. (I, p. 154)

The nature and desireability of perfection are questioned through central characters in the novels who begin to realize that most of Wraeththu's perfection is illusory: 'What had really changed since the first Wraeththu had come into the world? One selfish, ignorant race had been exchanged for another, more powerful, selfish, ignorant race' (I, p. 144). Dominant and subcultural values are placed in ironic juxtaposition as the central question of the trilogy – are the Wraeththu any better than mankind? – is pursued.

Further, by depicting the ultimately morally dubious Gelaming tribe, and their manipulative leader, Thiede, as the aspiring elite of the Wraeththu, Constantine is careful to show that former subordinate groups may become hegemonic in a way which reintroduces inequalities and hierarchies. The Gelaming's questionable assumption of dominance and superiority is thus hinted at in their style, which is too sanitised, too clinical, as Cal's response to their city implies:

> Immanion shone far beyond my dreams. We docked in the morning, stepping onto a harbour of sparkling mica. It was so clean. Unbelievably, shockingly clean. The brightness made my eyes ache. (III, p. 354)

The Gelaming function as the new orthodoxy to be resisted, and in a final redistribution of power brought about by Cal which shakes

Gelaming authority forever, Immanion undergoes a crucial style overhaul:

> On the highest levels, the stone still shone, and there were wide avenues where, in the morning, the light would dance. But now, there were also places where the light would never reach, the dark alleys, the subterranean canals and thoroughfares, where rats would creep and moaning ghosts disturb travellers from the lighter places above. (III, p. 369)

The Gelaming now appear to have their own subcultures, their disaffected, streetwise 'rats' and 'ghosts' who challenge the hegemony of Gelaming ideas and introduce an element of raw style into Gelaming perfection.

In drawing on a diverse range of subcultural styles to construct her post-apocalyptic world, Constantine employs the familiar late-twentieth-century literary technique of *bricolage*, producing a number of hybrid fictional societies held up for the reader's enjoyment and – ultimately – judgement. Plundering 'real-world' subcultures for the depiction of imaginary 'text-world' lifestyles also mimics the fundamental mechanism of *theft* at work in actual subcultural stylistic formations. The most important trends in youth style since the 60s have been generated by a rehabilitation and raiding of previous sartorial styles, and there has been a general trend in contemporary British culture which ransacks various historical moments for their key stylistic expressions and then re-inserts and recombines them in current fashion.[21] Constantine's literary technique mirrors this process and by doing so implicitly legitimates and demarginalises the mechanisms of creativity at work in British youth subcultures. It also, incidentally, means that the trilogy has gained a certain cult status amongst its British readers.

Constantine implies that there is a *mutational* relationship between mainstream cultures and subcultures. Mutation has been a favourite motif of post-Holocaust fiction, notably in John Wyndham's *The Chrysalids* (1955) and Walter M. Miller's *A Canticle For Leibowitz* (1960), and Constantine exploits this familiar theme to reinforce the sense that Wraeththu culture, and biology, retain aspects of the old (humanity) as well as developing elements of the new. This device primarily serves her message that humanity can and must change, particularly in the field of gender categories and sexual behaviour. But it also relates to the debate about the value of *bricolage* and the plundering of past styles so characteristic of post-war British subcultures. Frederic Jameson has criticised what he sees as a 'mass flight into nostalgia' represented by the recirculation of old images and styles. According to Jameson, loss of faith in the future has produced a culture which can only look backwards and re-examine key moments of its own recent history with a sentimental gloss and a soft-focus lens.[22] Whilst it should be said that Constantine's depiction of a post-apocalyptic world is imbued with a certain romanticism generated by its fantasy elements (myth, magic, medievalism), it does nevertheless raise the question of the relationship to the past, and to pre-existing ideologies, as a problem rather than a given. Hence we are constantly reminded that Wraeththu may have supplanted humanity, but they constantly benefit from that which man has left behind: their cities, their cars, even their religious festivals. The novel implies that subcultural difference is based on mutation, not unique difference, and warns against a too-easy assumption that the subculture subverts the ideologies of the norm.

The motif of mutation also reflects the typical characterisation of subcultures by the mainstream as freakish, unnatural and monstrous. In her depiction of the Wraeththu mutants as highly aestheticised potential superbeings, Constantine would no doubt concur with Alfred Jarry's assertion that 'It is conventional to call 'monster' any blending of dissonant elements [...] I call 'monster' every original, inexhaustible beauty.'[23] Most suggestively, Hebdige argues that subcultural styles should be regarded as *'meaningful* mutations' which, in counterposition to mainstream cultural styles, 'assume monstrous and unnatural features'.[24]

The motif of monstrous mutation has a further set of associations if we consider that the trilogy is rife with vampire imagery. When Pell first kisses Cal, he says, 'I expected a vast vampiric drain on my lungs' (I, p. 18) and one of the apocryphal myths told about the Wraeththu by fearful humans is that they 'they drank blood' (I, p. 11). Later when Pell encounters Cobweb, he says, 'I thought of a vampire face and hollow eyes' (I, p. 164). Swift's experience of a night of rough sex with Cal leaves him feeling like a vampire's victim:

> He murmured as if in pain, fretfully, then his arm lashed back and hit the pillow. His eyes were blazing, I reared up to escape them, but he caught hold of me, so strong, lifting himself. His mouth found my neck; he wanted blood [...] In the morning I felt

> as if I'd been fighting for my life all night [...] He was gorged on my vitality. I could hardly move. (II, p. 164)

In Book 3 Cal takes a ritualised form of vampiric aruna (sex) with a Wraeththu from the Elhmen tribe (III, pp. 214–16), and like all vampires, Cal is wary of mirrors: 'Someone gave me a mirror and I saw the future' (I, p. 157). The myth takes on a form of concrete reality when we find that the warlike leaders of the Varrs, Terzian and Ponclast, have actually drunk the blood of their victims:

> From surviving victims, we have discovered that the Varrish elite feast upon the flesh of their own kind, the most prized vintages in their wine cellars being barrels of Wraeththu blood. (II, p. 317)

Most importantly, Harhune, or inception, the method by which men become Wraeththu, involves a blood transfusion which mutates man into Har. Vampiric style is well-beloved by Goths, but it is also central to popular fears about gay male sexuality. Thus by using such vampiric imagery, Constantine alludes to the common association of gay men with vampires, what Ellis Hanson describes as:

> ... essentialist representations of gay men as vampiric: as sexually exotic, alien, unnatural, oral, anal, compulsive, violent, protean, polymorphic, polyvocal, polysemous, invisible, soulless, transient, superhumanly mobile, infectious, murderous, suicidal, and a threat to wife, children, home, and phallus.[25]

Queer theory has argued that to vampirise the gay man, particularly the gay man with AIDS, is to relegate him to the space of the abject, the site of the unnameable and unspeakable. Hanson says:

> Typically, in media representations of AIDS, I find neither people who are living with AIDS nor people who have died with AIDS. What I find, rather, are the spectacular images of the abject, the dead who dare to speak and sin and walk abroad, the undead with AIDS. I find a late-Victorian vampirism at work, not only in media constructions of AIDS now, but in the various archaic conceptions of same-sex desire which inform the present 'Face of AIDS'.[26]

By making her beautiful, 'fey yet masculine' Wraeththu-kind a product of a form of vampiric blood transfusion, Constantine presents her fictional mutants as a rehabilitated, idealised form of revenant. They are the undead who survive the death-like experience of inception and rise after three days of abjection as reborn superbeings: 'within three days of being infected with Wraeththu blood, the convert's body has completed the necessary changes' (I, p. 50). The damaging myth of the gay male vampire is transformed by transfusing drafts of romance and fantasy: gay male sex is depicted not as abject but as exquisite, and the apocalyptic 'gay plague' of AIDS becomes the redemptive 'infection' that is the spread of Wraeththukind and their superior ways.

Current theories of the 'gay gaze' also appear to inform Constantine's trilogy. Volume one is dedicated 'to the almond eyes', and while this evokes certain punk and Goth styles of makeup, it also signals a central concern with the eye. Indeed, because Wraeththu culture is a highly visualised one, with an emphasis on visual style, there are many scenes where Wraeththu are depicted as engaged in unabashed, reciprocal *looking*. We are told that 'Wraeththu are never ashamed to admit they are beautiful' (I, p. 23) and aruna between Wraeththu always involves a complex and erotic interplay of gazes:

> He stood up and carelessly pulled off his shirt. His skin was dark, his hair almost white. I looked away and after a while he said, 'Oh Swift, can't you bear to look at me?'
> 'Moswell said it would be indelicate to stare.'
> 'Nonsense, it turns me on. I want you to admire me ...'
> We went to the windowseat and looked out at the garden and it felt wild and magical to be naked. Anyone might have looked up and seen us ...
> 'Look at yourself,' he told me. (II, pp. 158–60)

Recent queer theorists have pointed out that assumptions that the gaze is always phallic and exploitative are grossly simplistic and leave no space to conceptualise queer relations of looking.[27] By depicting the erotic gay male gaze, Constantine acknowledges the possibility of perverse and enjoyable relations of looking, as opposed to the exploitative male gaze. Caroline Evans and Lorraine Gamman have pointed out that 'when individuals cruise each other on the street, or in clubs, the mutual exchange of glances is sexualised and often reciprocal'.[28] Wraeththu cruise continually, due to the huge importance of aruna in their largely non-monogamous society, and because physical beauty is the norm and frankly admired. Everyday Wraeththu looking does not connote voyeurism or exploitation, but a mutual appreciation of inherent worth. Thus Constantine uses the trilogy as a fictional space in which to conceptualise queer relations of looking. Women's traditional exhibitionist role, whereby they are similarly looked at and displayed, is taken on by *all* Wraeththu in mutual exchanges. Early theories of the gaze argued that 'the male figure cannot bear the burden of sexual objectification',[29] but over the last twenty years the male body has been increasingly displayed and sexualised. Constantine responds to this trend and depicts a homoerotic looking at the male body (albeit somewhat mutated!) which invites the participation of the reader, be they male or female. As Suzanne Moore argues, 'Homoerotic representations, far from excluding the (voyeuristic) female gaze, may actually invite it.'[30]

However, Constantine is careful to remind us that queer looking can still turn into exploitative male gazing. Pell wears a talisman depicting a 'sacred eye' whose traditional associations of vision and wisdom are gradually supplemented by more sinister aspects: the possibility of surveillance as a means of control. In a tense scene of conflicting gazes, Pell realises that the Gelaming leader Thiede has corrupted Wraeththu looking into what we can recognise as Foucauldian surveillance: [31]

> 'Where have you seen me before?'
> He turned back to the packing. 'Everywhere Pellaz, everywhere. I have seen through Thiede's eyes ...'
> All the chill came back to my flesh; my hand curled around Orien's talisman. Thiede's eyes; my life a spectacle. I was staring at a heavy pewter jug that stood on a table by my bed. I was thinking of the weight of it in my hands and the impact of it against the back of Vaysh's bent head. I was thinking of me, fleeing the tower and running just anywhere; all of this. Luckily, I was not thinking hard enough.
> Vaysh stood up. 'We must leave,' he said. 'Are you ready?'
> We looked at each other without liking. He knew that I had the power, even the desire to kill him, but he also knew just what had made Thiede choose me. I closed my eyes so I did not have to look at him. (I, pp. 214–15)

In this context, we can read Thiede's repulsion of the usual Wraeththu look – we are told, 'Thiede is very hard to look at directly, because his beauty is so alien and stark' (I, p. 257) – as a strong hint of his dangerous hegemonic ambitions.

All of Constantine's novels to date deal with gender issues, including *The Monstrous Regiment* (1990), which depicts a colony world where rampant feminism has produced a mad ruler – the Dominatrix – who plans to confine the men and treat them solely as sperm producers, and *Hermetech* (1991), in which sex provides the power to save an ecologically stricken Earth. As we have already seen, the *Wraeththu* trilogy evokes subcultural androgyny and revises myths of gay male sexuality, but in the final analysis Constantine's troubling of gender categories, or what June L. Reich (amongst others) has called 'genderfuck', is only a partial one. Reich describes 'genderfuck' as:

> ... the effect of unstable signifying practices in a libidinal economy of multiple sexualities [...] This process is the destabilisation of gender as an analytical category, though it is not, necessarily, the signal of the end of gender [...] the play of masculinity and femininity on the body [...] subverts the possibility of possessing a unified subject position.[32]

The trilogy certainly does not depict the 'end of gender'. In fact, conventional sexual roles ('soume'/submissive and 'ouana'/dominant) are retained, and a rather hackneyed – and, as many feminists have pointed out, flawed – notion of androgyny forms the basis of Wraeththu gender identity.[33] Constantine's conceptualisation of Wraeththu androgyny as a union of masculine and feminine traits fails to deconstruct those very conventional gender binaries. However, the supposed perfection of Wraeththu genderblending is deliberately placed under question by making their adoring new convert, Pell, a clearly naive and fallible narrator. And it doesn't take long for even this willing admirer to question the apparent marginalisation of the female in Wraeththu sexuality: 'Wraeththu combined the sexes by favouring the male. There are too many issues unraised, too many uncomfortable questions unanswered' (I, p. 154). The early glibness with which women have been excluded from the Wraeththu's post-apocalyptic vision of a superior world ('They that joined us [i.e. men], the lucky ones, will be the only survivors' (I, p. 74)) is later challenged through depictions of open debate:

> 'Perhaps every man on this earth could be incepted to Wraeththu – even the women! Does anyone really know? Has anyone ever tried to incept a full-grown man, or a woman?'
> Cal looked at me with distaste. 'You can be quite grotesque sometimes Pell,' he decided.
> To my surprise Cobweb agreed with me. 'No, Cal,' he said. 'It is grotesque to think otherwise. That is man's smallmindedness; man's fear of questioning important issues.' (I, p. 141)

The seeds of dissatisfaction are sown in the reader via depictions of Pell's human friend, Kate, as clearly prime Wraeththu material, yet prevented by her sex from joining the elite group: 'Boyish in her manner, barely older than myself, she sprawled on a stool like an ungainly colt, appraising me with green eyes. "I curse the day I was born a woman"' (I, p. 84). Thus Constantine encourages a resistant reading of the Wraeththu's supposed androgynous perfection. Their genderblending is not such an inevitable 'genderfuck' as we are lead at first to believe. To reinforce this message, Constantine challenges the fundamental binarism of Wraeththu androgyny by concluding her trilogy with an image of power and perfection based on a perversely multiple sexual grouping, a gay male *ménage à trois*. Pell, Cal and Caerue join forces in a sexual and spiritual union more powerful than any of the binary unions that have gone before, and a sexual threesome becomes, in a camp parody of the Trinity, 'the earthly Triad' (III, p. 388).

Nevertheless, it must be said that whilst Constantine goes some way to subverting conventional myths of gay male sexuality, and challenging the efficacy of subcultural styles of androgyny as a form of resistance to gender norms, she entirely neglects lesbianism as a potentially subversive subcultural lifestyle. Constantine chooses to wait until the end of the final novel to reveal the existence of female counterparts (indeed superiors) to the Wraeththu, the Kamagrians. They remain desexualised (there is no suggestion of lesbian desire in the trilogy) in a way which the Wraeththu emphatically do not. And because they are finally depicted via the rather misogynist perspective of Cal, who has been functioning as the narrator of the final volume, they remain – as women have been traditionally – mysterious and rather threatening Others, not just for Cal but for the reader also. Most damaging in the context of Constantine's commitment to the subversive aspects of style, the Kamagrians appear to entirely lack a distinctive style or aesthetic sense. Ultimately, Constantine's fictional exploration of the subversive potentials of subcultures neglects female subcultures, particularly lesbian ones. She reinforces the mainstream stereotype of youth subcultures as predominantly male, and unfortunately rather too literally reiterates the dictum that 'style maketh the *man*'.

By portraying the Wraeththu as ambiguous agents of humanity's end, Constantine fictionally literalises the apocalyptic nuances of British subcultural styles. But she also provides a context in which to question the subversive values of those styles, and show that subcultural resistance may not always bring hegemonic values to an end. Her fictional depiction of 'mankind's funeral' functions less as a grim warning than as a reminder that apocalypse is never absolute.

Notes

1. All citations to the *Wraeththu* trilogy refer to the following editions: Storm Constantine, *The Enchantments of Flesh and Spirit: The First Book of Wraeththu* (New York: Tor Fantasy, 1987); *The Bewitchments of Love and Hate: The Second Book of Wraeththu* (New York: Tor Fantasy, 1988) and *The Fulfilments of Fate and Desire: The Third Book of Wraeththu* (New York: Tor Fantasy, 1988).
2. For example, Doris Lessing, Suzy McKee Charnas, Sally Miller Gearhart.
3. There have been a number of post-holocaust trilogies, including Richard Cowper's *Corlay* trilogy (1976–82) and William Barnell's *Blessing* trilogy (1980–1). However, it is the trilogy of novels by Suzy McKee Charnas (*Walk to the End of the World* (1974), *Motherlines* (1978) and *The Furies* (1994)) which stands most starkly in comparison to the *Wraeththu* novels, in its lesbian feminist analysis of gender issues.
4. Paul Willis, *Common Culture* (Milton Keynes: Open University Press, 1990), p. 85.
5. Ibid. p. 128.
6. Dick Hebdige, *Subculture: The Meaning of Style* (London: Routledge, 1988), p. 17.
7. Ibid. p. 3.
8. Ibid. p. 18.
9. Quoted in Hebdige, p. 100.
10. Simon Frith, 'Only Dancing: David Bowie Flirts With the Issues', in *Zoot Suits and Second-Hand Dresses: An Anthology of Fashion and Music*, ed. Angela McRobbie (London: Macmillan, 1989), p. 135.
11. Ibid. p. 134.
12. John Clute and Peter Nicholls (eds.), *Encyclopedia of Science Fiction* (London: Orbit, 1993), p. 260.
13. David Laing, *One Chord Wonders: Power and Meaning in Punk Rock* (Milton Keynes: Open University Press, 1985), pp. 117–18.

14. McRobbie, p. 42.
15. See Laing p. 40 for a full discussion of the history of the word 'punk' and its connotations of male homosexuality.
16. Hebdige, p. 24.
17. Ibid. p. 27.
18. Willis, p. 126.
19. Laing, p. 24.
20. Hebdige, p. 27.
21. Willis, p. 88.
22. Frederick Jameson, 'Postmodernism, the Cultural Logic of Capital', in *Postmodern Culture*, ed. H. Foster (London: Pluto Press, 1985), p. 64.
23. Quoted in Hebdige, p. 102.
24. Hebdige, p. 131.
25. Ellis Hanson, 'Undead', in *Inside/Out: Lesbian Theories, Gay Theories*, ed. Diana Fuss (London: Routledge, 1991), p. 325.
26. Ibid. p. 325.
27. For example, Elizabeth Grosz says; 'Many feminists ... have conflated the look with the gaze, mistaking a perceptual mode with a mode of desire. When they state baldly that 'vision' is male, the look is masculine, or the visual is a phallocentric mode of perception, these feminists confuse a perceptual facility open to both sexes ... with sexually coded positions of desire within visual (or any other perceptual) functions ... vision is not, cannot be, masculine ... rather, certain ways of using vision (for example, to objectify) may confirm and help produce patriarchal power relations.' Elizabeth Grosz, 'Voyeurism/Exhibitionism/the Gaze,' in *Feminism and Psychoanalysis: A Critical Dictionary* ed. Elizabeth Wright, (Oxford: Blackwell, 1992), p. 449.
28. Caroline Evans and Lorraine Gamman, 'The Gaze Revisited, Or Reviewing Queer Viewing', in *A Queer Romance: Lesbians, Gay Men & Popular Culture*, eds. Paul Burston and Colin Richardson (London: Routledge, 1995), p. 15.
29. Ibid. p. 30.
30. Lorraine Gamman and Merja Makinen, *Female Fetishism: A New Look* (London: Lawrence & Wishart, 1994), p. 55.
31. Michel Foucault, *Discipline and Punish: The Birth of the Prison*, trans. Alan Sheridan (Harmondsworth: Peregrine Books, 1979). See also Foucault's discussion of the panopticon in Michel Foucault, *Power/Knowledge: Selected Interviews and Other Writings 1972–77*, ed. Colin Gordon (Brighton: Harvester, 1980), pp. 146–65.
32. June L. Reich, 'Genderfuck: The Law of the Dildo', in *Discourse*, 15: 1 (Fall 1992), p. 125.
33. For a concise discussion of the issues raised by the notion of androgyny, see Toril Moi, *Sexual/Textual Politics: Feminist Literary Theory* (London: Methuen, 1985), pp. 13–15.

14

Jews and *Independence Day*, Women and Independence Day: Science Fiction Apocalypse Now Evokes Feminism and Nazism

MARLEEN BARR

Kai Erikson describes what he calls 'a new species of trouble': 'Indeed, everything out there can seem unreliable and fearsome [...] [T]he ground itself can no longer be relied on [...] People feel that something noxious is closing in on them, drifting down from above, creeping up from underneath, edging in sideways, fouling the very air and insinuating itself in all the objects and spaces that make up their surroundings [...] The point is not that a particular region is now spoiled but that the whole world has been revealed as a place of danger and numbing uncertainty.'[1] The new species of trouble I will discuss applies to women in general and Germans in particular and results from paradigm shift, not the haphazard disasters Erikson describes (such as nuclear accidents and floods). I am concerned with women's response to the aftermath of Betty Friedan's *The Feminine Mystique* and Germans' response to the aftermath of Second-World-War-era Nazism. Women and Germans face a new species of trouble: fallout emanating from changes in old, abhorrent rules. Sexism and Nazism, which were once socially acceptable, have become something noxious, able respectively to close in on women and Germans everywhere in the world.

Although new generations of Germans are innocent of Nazi atrocities once championed by the German government, they feel guilty about them. Although new generations of women are highly sensitized to sexism once championed by all patriarchal institutions, they feel vulnerable to existing sexism they cannot nullify.

For Germans and women, ignorance *was* bliss. Before Friedan, women had no means to articulate the problem which had no name. Before the War's conclusion, Germans had a means to say that they did not know. Germans and women, now lacking such comfort, face the world in the manner Erikson describes.[2] For women, the whole world (their personal and professional lives) has been revealed as a potential place of dangerous sexism which can appear at any moment. For Germans, the whole world (their German soil and international interactions) has been revealed as a potential place of embarrassing past history which can appear at any moment.

Walter Abish's *How German Is It* exemplifies why Germans cannot rely upon their own ground. He describes Second-World-War-era corpses emerging from a hole which unexpectedly appears in front of a bakery in a German town. The corpses creep up from underneath and foul the town's air. Sink holes which routinely appear in Florida are innocuous. Not so for the mass graves routinely discovered in Germany. Not so for the surprise hole, the new species of trouble Abish's German protagonists face. Innocent, young and middle-aged Germans, when they venture abroad, never know when they will be called Nazis and shunned. Nor can Germans relax in their own country. J. S. Marcus, the Jewish author of *The Captain's Fire*, the first American novel about Germany written after the fall of the Berlin Wall, has his Jewish-American protagonist explain that trying not to think about Nazis in Germany is analogous to trying not to think about sex in a porno shop. 'Why go in at all?' Marcus' protagonist asks.[3]

Daniel J. Goldhagen's *Hitler's Willing Executioners* has closed in upon Germans. The *New York Times* describes the impact Goldhagen's September 1996 book tour had in Germany: 'Goldhagen's book has left ordinary Germans to grapple with his conclusion that it was something evil in their very nature – not just the manipulation of Nazism – that led to genocide.' 'We should revise the picture of automatons following orders without free will,' Mr Goldhagen said in opening remarks that kicked off two hours of bitter discussion followed by a separate television panel debate marked by the same hostility. 'It became apparent that many Germans are not prepared to accept a picture of their forebears that does not take into account the particular historic circumstances [...] Throughout the postwar era, one of the most difficult questions for Germans to answer has always been why it was they – and not other anti-Semitic Europeans – who took it upon themselves to eliminate six million Jews.'[4] Goldhagen defended his ideas by asking 'Does anyone here [...] believe that people who were killing Jews did not have a view of what they were doing?' (Cowell 4). Goldhagen is absolutely correct. I must say too that I feel sympathy for new generations, innocent Germans, who must grapple with their relationship to the Holocaust generation.

Sexism now impacts upon women similarly to Nazism's impact upon contemporary Germans: sexism potentially exists in all the spaces which make up women's surroundings. A woman, at any moment, can be turned into an object. Just as a law-abiding black can be arrested for suspected trespassing when present in an affluent American white neighborhood, a law-abiding woman walking in any American hotel lobby can be arrested as a suspected prostitute. Sexism and accusations of Nazism are noxious and can unexpectedly close in at any time. Being a woman or a German navigating everyday life is analogous to trying to take a relaxing ocean swim after watching *Jaws*. You know that a fearsome something is out there and you cannot protect yourself when it decides to get you. For my hypothetical swimmer – as well as for Germans and women – the whole world has been revealed as a place of danger and numbing uncertainty. Women and Germans now face a new species of *Tsuris* (the Yiddish word for trouble – which young Americans would translate as 'shit hits the fan') previous generations did not experience.

It is logical, then, for women and Germans to behave according to Erikson's description of disaster victims. He explains that these victims 'are very apt to develop a sense of being out of control, of being caught up in forces that capture them and take them over. Survivors of severe disasters experience not just a sense of vulnerability but a feeling of having lost a certain immunity to misfortune, a feeling, even, that something terrible is bound to happen.'[5] Susan Sontag argues that science fiction about apocalypse now provides a safety valve for these experiences.

Sontag explains: 'What I am suggesting is that the imagery of disaster in science fiction is above all the emblem of an inadequate response [...] the inadequacy of most people's response to the unassimilable terrors that infect their consciousness. The interest of the films [...] consists in this intersection between a naive and largely debased commercial art product and the most profound dilemma of the contemporary situation.' For women and Germans, sexism and Nazism are respectively a most profound dilemma of the contemporary situation.[6] Although this dilemma – the fear of the unexpected – causes widespread psychological pain, efforts to assuage this pain are inadequate. For most Americans, juxtaposing Nazism with all Germans remains a knee-jerk response. Most sexist people remain unrepentant (and Anita Hill's naming of sexual harassment is a recent occurrence). The imagery of disaster in science fiction about apocalypse is an emblem of efforts to address such profound contemporary dilemmas.

I will discuss *Independence Day*, an apocalyptic science fiction film directed and co-written by a German (Roland Emmerich) as well as five apocalyptic science-fiction novels written by women since 1990: Eleanor Arnason's *Ring of Swords*, Margaret Wander Bonanno's *The Others*, Suzy McKee Charnas' *The Furies*, Michaela Roessner's *Vanishing Point*, and Joan Slonczewski's *Daughter Of Elysium*. My comments pertain to newness. All Germans now live under new democratic rules. Rules regarding the terminating of sexism are still being formulated. Women writers of apocalyptic science fiction evoke Nazism to deal with the anxiety this textual lack causes. Emmerich's film reflects his desire to separate himself from his country's Nazi past. The women writers imbue their texts with allusions to Nazism to underscore that sexism is heinous too.

Sexism impacts upon women and Nazism impacts upon Germans in a manner analogous to Sontag's comment about the bomb. She says: 'It became clear that, from now on to the end of human history, every person would spend his individual life under the threat not only of individual death, which is certain, but of something almost insupportable psychologically – collective incineration and extinction which could come at any time, virtually without warning.'[7] Now that Friedan has named the problem that has no name, all educated women live with the knowledge that they have nowhere to go to escape patriarchy's ability to cause personal and professional identity extinction. Now that no German can claim to be ignorant about the incineration of people some Germans instituted, no German can escape from recent history. To deal with these contemporary dilemmas, Arnason, Bonanno, Charnas, Emmerich, Roessner, and Slonczewski look to apocalyptic science fiction. They tell stories of collective incineration and extinction which can come at any time, virtually without warning, and can affect everyone. For these artists, apocalyptic science fiction provides a means to comment upon an accident of birth, being born German or female.

INDEPENDENCE DAY: EMMERICH'S DAY OF ATONEMENT FOR GERMANS

Mona Chang, protagonist of Gish Jen's *Mona In The Promised Land* – the story of Chinese-American Mona's decision to convert to Judaism – asks this question: 'How can everybody in the fucking world be Jewish?' Emmerich's *Independence Day* answers her question.

Mona asserts that 'Jewish is American'.[8] So is science fiction. So is Independence Day. Emmerich turns to American culture, to science fiction and the Fourth of July, to contend with Germany's recent past. He approaches the near-annihilation of Jews by situating his story in the place of their greatest proliferation – in America. Hence it is no accident that this German's blockbuster science fiction film takes the quintessentially American holiday as its title. *New Yorker* critic James Wolcott, however, states that the film is merely self-referential: what 'makes the movie "meta" is not only its self-awareness but the fact that it doesn't mean anything beyond itself [...] INDEPENDENCE DAY has no subtext. It's all *Ubertext*.'[9] I beg to differ. *Independence Day* reflects a German director/writer's desire to position Americans – and all people on Earth – as victorious Jewish freedom fighters.

Emmerich, now 40, grew up in a German world where virtually no-one was Jewish. Although he and his generation are innocent, as Germans they experience guilt and hostility. Paul Verhoeven, director of *Robocop* and *Total Recall*, says that the 'U.S. is desperately in search of an enemy [...] The Communists were the enemy, and the Nazis before them, and now that wonderful enemy everyone can fight has been lost.'[10] But the Nazis are *still* the enemy of those young and middle-aged Germans who are embarrassed by their country's past. Largely due to Hollywood films' prevalent stereotype that all Germans were and are Nazis, all Germans are associated with Nazism. In *Independence Day*, Emmerich confronts such associations by presenting seemingly invincible extraterrestrials – entities who are even more devastating than Nazis. *Independence Day*, in which Jews become Everyman and a Jew and a black win the day, is Emmerich's day of atonement for a cataclysm which at once has everything (because he is German) and nothing (because he was born after the war) to do with him. This point becomes clear when Emmerich links his career to his childhood: 'For me [...] going on a science-fiction movie set is like visiting toyland. You see, my brother trashed all my toys when I was a kid. It's very

Freudian. For my movies you can blame my brother Andy.'[11] While growing up in 1950s Germany, Emmerich experienced a disgraced country of trashed toys: no more visions of world domination, no more V-2 rockets, no more Fatherland, no more Hitler rallies. Since he cannot be proud of his country, he hits back by 'trashing' monstrous extraterrestrial meta-Nazis.

Sontag seems to allude to Emmerich's psychological motivations. She says: 'In the face of the monster from outer space, the freakish, the ugly, and the predatory all converge – and provide a fantasy target for righteous bellicosity to discharge itself, and for the aesthetic enjoyment of suffering and disaster.' Emmerich's Nazi monsters from outer space are a revenge fantasy target for him. He enjoys the fact that despite the aliens/Nazis' power, despite their ability to holocaust the entire earth he can, in the end, make them experience suffering and disaster. Sontag continues: 'alongside the hopeful fantasy of moral simplification and international unity embodied in the science fiction films lurk the deepest anxieties about contemporary existence [...] the science fiction films reflect powerful anxieties about the condition of the individual psyche.'[12] In contemporary Germany – a rich country with a social safety net – the deepest anxiety of contemporary existence, the most powerful collective anxiety about the condition of the individual psyche, is association with the Holocaust. For Emmerich's generation to say 'I didn't do it' is moral simplification; their parents' generation *did* do it. Guilt lurks above Emmerich's generation, ominously hovering like his film's spaceships. *Independence Day* is Emmerich's fantasy about independence from guilt, eradicating guilt, winning the day against an even more monstrous version of the Nazi past. '[F]antasy can [...] normalize what is psychologically unbearable, thereby inuring us to it. In one case, fantasy beautifies the world. In the other, it neutralizes it.'[13] *Independence Day* normalizes being German, normalizes inheriting the Holocaust. *Independence Day* beautifies the German landscape littered with broken toys, defeats the monster, puts Humpty Dumpty back together again.

Independence Day neutralizes Nazism. The dead aliens/Nazis are the film's *raison d'être*; the millions of human deaths the aliens cause do not matter. Wolcott explains: 'At the end of *Independence Day* millions are dead, entire cities are destroyed [...] and the battered survivors stand around grinning as if they'd just won a volleyball game. They couldn't be more gee-whiz. There's no weight to what happened – no hint of hard, scavenger times ahead. Only Hollywood in the nineties could arrange such a pain-free, upbeat Armageddon.'[14] Winning volleyball games and gee-whiz attitudes belong to childhood – and Emmerich is trying to put his broken toys back together. In order to do so, he makes the aliens' demise, not human demise, most important. The millions of deaths (more people than Hitler killed) and the destroyed cities (more devastation than Hitler caused) are a gee-whiz way of fantasizing about terminating fictitious Nazis (who are more powerful than real Nazis). There is a precursor to Hollywood's pain-free upbeat 90s Armageddon: America's pain-free upbeat 40s Armageddon. Bombing Hiroshima and Nagasaki was, according to Americans, a good thing. According to Nazis, killing Jews was a good thing too. For Nazis, the Holocaust was a pain-free, upbeat Armageddon. Nazis were not impacted upon by mass Jewish death; Americans were not impacted upon by mass Japanese death; Emmerich, his characters, and his audience are not impacted upon by the mass human death *Independence Day* portrays.

Will American science fiction film makers try to come to terms with Hiroshima and Nagasaki? Probably not. Science fiction 'films reflect world-wide anxieties, and they serve to allay them'.[15] Germans are steeped in guilt about the Holocaust and world-wide anxiety is directed toward them. Americans, in contrast, are not guilty about dropping the bomb; the world does not chastise them for having done so.

In the manner of unheeded warnings about the impending Holocaust, no-one listens to Russell Casse, Emmerich's crop dusting California desert denizen who claims that he was kidnapped and experimented upon by aliens. This goy incarnate is a Jew – a survivor of experiments Nazi doctors/aliens performed upon him. His capture portends that Nazi/aliens will come to earth to exterminate all humans/Jews. WASP/Jewish President Thomas J. Whitmore announces 'we are being exterminated'. 'We' means all the humans on earth. Emmerich, the German, creates a fantasy in which humans, who are all recast as Jews, win the day.

Audiences, of course, realize that in *Independence Day* blacks are for once portrayed as forming an intact nuclear family and the Jewish man wins the prize, the gentile woman (Whitmore's communications director Constance Spanos). It is also obvious that the Jew and the black (whom I position as another Jew) save the world. As Wolcott observes, 'When [Will] Smith [who plays marine Captain Steven Hiller] and [Jeff] Goldblum [who plays David

Levinson] share a spacecraft on a near-suicide mission to the alien mother ship, wisecracking in the face of extinction, they show that the soldier and the scientist – and, more important, the black and the Jew – can bond; they're the Cornel West and Michael Lerner of extraterrestrial *Tikkun*.[16] In *Jews and Blacks*, Lerner states that Jews will be targeted by an American fascistic movement which could become a viable entity in twenty to thirty years.[17] *Independence Day* responds to this possibility by making everyone vulnerable to holocaust and positioning Jewish expertise as the catalyst for global salvation.

The film's Jewish/black dynamic duo can take action because of the word of a Jewish father who behaves like a Jewish mother. David Levinson becomes aware that he can spread a computer virus to the aliens immediately after his father Julius warns him to avoid catching cold. 'Cold' is the word as savior. A word for defining 'virus' saves the world because of a Jewish father and son, people Nazis defined as virulent intruders.

Whitmore speaks in a manner appropriate for a Warsaw ghetto partisan or an Israeli: 'We fight for the right to exist. We will survive,' he says. Humanity survives because a black and a Jew, while under alien/Nazi attack, are both Jews who defy two American myths: that blacks and Jews should be at odds, and that black men and Jewish men do not have the right stuff to qualify as macho men. Counter to Wolcott's aforementioned opinion, Emmerich's alien encounter story is not simplistic. It concerns encounters between those who are often alien to each other: blacks and Jews, and Jews and gentiles. Humanity survives because a Jew and a black pool their talents – and Casse, the abducted goy, becomes the sacrificial lamb.

When Emmerich casts humanity as Jews he creates new Jews – Jews who fight back and vanquish aliens/Nazis. Whitmore directly tells an alien that he wants peace. The alien is very uncooperative; it says that it wants humans to die. Instead of failing to take defensive action, instead of meekly going to the slaughter, Whitmore tries to defend the concentration camp for humans/Jews the entire earth has become. 'Nuke the bastards,' says Whitmore. The aliens cause nuclear missiles, those all-pervasive symbols of American manhood, to become impotent. A new definition of American manhood wins the day: cooperation between Others, a Jew and a black.

Independence Day opens with an image of a fractured text, a signal that the film will blast apart stereotypical masternarratives relating to race and ethnicity. In America, a good alien – Superman (who is Clark Kent from Smallville, not, for example, Irving Horowitz from the Lower East Side) – is a hero. In Emmerich's America, a Jew becomes a heroic Superman. Julius Levinson, the Jewish father/mother, announces this fact. 'Without my David you would be dead now,' he says. Earth is saved by a triumvirate more appropriately American than the Father, Son, and Holy Ghost: the Jew, the black, and the sacrificial redneck. This triumvirate, a new American alliance envisioned by a German, functions as the messiah who saves a world in which all humanity is the Jew.

Emmerich's film, his day of atonement for Germans which is a close encounter with a Nazi/alien third kind, brings another Jewish holiday to mind: Hanukkah. Like the story of the lamp which somehow burns, Emmerich's story is about a miracle. This miracle entails the fact that alien ships which holocaust earth's cities can ultimately be burned by humans/Jews. Emmerich portrays horrors Nazis never perpetrated. Nazis did not (and could not) engulf the White House in flames and topple American icons. Unlike real Nazis, Emmerich's Nazis/aliens can burn the entire world. But: the Hanukkah lamp burns – and so do the alien ships. This miracle is important. Extraterrestrial invasion, after all, might be as real as Nazi invasion. Emmerich calls the possibility that the U.S. government is harboring dead aliens found near Roswell, New Mexico this generation's 'own mythology' (Corliss 59). The Nazi/aliens could already have arrived. Extraterrestrial induced apocalypse could unite the world. During the summer of 1996 flames did symbolize both a united and disunited world. The Olympic flame burned in Atlanta; TWA flight 800 burned in the Atlantic. Which burning image will prevail?

INDEPENDENCE DAY AND FEMINIST PYROTECHNICS

How can a marine navigate an alien spaceship? How can a computer maven introduce a virus to an alien computer system? Gish Jen offers a good answer: 'In outer space there are no rules.'[18] Outer space, then, provides a blank page upon which women can rewrite the rules which position them as the Other, vulnerable to pervasive sexism. I read another of Erikson's comments in terms of these rules. He says that the 'second thing to be said about these new troubles is that they involve toxins: they contaminate rather than merely damage; they pollute, befoul, and taint rather than just create wreckage; they penetrate human tissue indirectly rather than wound the surface by assault of a more straightforward kind.'[19] Patriarchy often defines women as toxic intruders whose presence befouls male bastions. Nazis certainly viewed Jews as those who taint, as a virus indirectly penetrating societal tissue. Perhaps identification with Jewish victimization (in relation to their post-Friedan awareness of Otherness) causes women science-fiction writers to allude to Nazis when making their apocalyptic worlds which 'become contaminated, burnt out, exhausted obsolete'.[20]

The novels I consider involve clashes between differing groups which could result either in the end of the world or a particular group's demise. Arnason creates antagonists who are humans and militaristic humanoid aliens; Bonanno creates antagonists who are primitives and technologically sophisticated island dwellers; Charnas creates antagonists who are women and men (good old garden-variety adversaries); Roessner creates antagonists whose theories about why most of humanity vanishes differ; and Slonczewski creates antagonists who are virtually immortal aliens and sentient machines. Each author describes protagonists' efforts to arrive at their version of Independence Day: freedom from either racism, loss of subjectivity, or total annihilation.

There, at first, seems to be no Independence Day in store for the protagonist of Bonanno's *The Others*. When we meet Lingri, chronicler of the Others, we learn that 'the People' cause her people (called 'the Others') to be 'on the verge of our annihilation'. The People's annihilation methods would be familiar to the German generation which precedes Emmerich's generation. According to the official orders of the People's Purist Praesidium called 'On the Disposal of the NonPeople', the people intend to 'contain, reduce, and eliminate' the Others. This elimination plan leads to a dispute: 'the dispute being whether actively to exterminate them [the Others], or simply to cordon them off and let them starve'.[21] Lingri lives among the People and learns not to hate them. Lingri, the potential victim of genocide, achieves freedom from despising those who are Other in relation to her. Her fellows achieve freedom too. They set sail from their native archipelago to islands in their planet's polar region: in other words, they become refugees. *The Others* tells a science-fiction apocalypse story in terms of Leon Uris' *Exodus*, resembling what H. G. Wells might have written if a time machine had enabled him to know about Nazis when he created Eloi and Morlocks.

Eleanor Arnason's *Ring of Swords* also presents one group (the *Hwarhath,* aliens who are either hairy militaristic males or matriarchs) deciding whether or not to define another group (humans) as people. The situation is complicated by the fact that the *Hwarhath* view heterosexuality as a perverted abomination. When the *Hwarhath* rail against the evils of heterosexuality, Arnason emphasizes that nothing is good or bad but discourse makes it so. The aliens' arguments against heterosexuality allow them to take very valid potshots at human male violence in particular and patriarchy in general. In fact, *Ring of Swords* focuses more upon gender war than upon potential apocalpytic war between the *Hwarhath* and the humans. Although they are obsessed with military precision, the hulking hairy male *Hwarhath* are wannabe feminist utopians. A *Hwarhath* male warrior, for example, never does physical harm to women and children. Arnason, then, at once emphasizes and nullifies Joanna Russ' observation that 'men are dangerous'.[22] When biological researcher Anna Perez becomes aware that a male *Hwarhath* is having sex with a human male translator (Nicholas Sanders), and when she learns about *Hwarhath* matriarchy, she encounters a situation analogous to Han Solo having sex with the Wookie before he heads toward the planet Ursula Le Guin describes in *The Word For World is Forest.*

When the *Hwarhath* try to decide whether or not humans should be classified as people (and hence spared the military attack which will cause humanity's destruction), Sanders is called 'an animal, a very clever one, able to mimic the behavior of a person'. Nazis, of course, described Jews as vermin, as animalistic imitators of humans. If interplanetary war occurs, the *Hwarhath* will fight the humans as if they were '[s]mall vermin. A destructive bug.' The humans, potential contaminators of the aliens' social tissue, are threatened with 'a final solution'. Luckily, the *Hwarhath* decide that humans are people rather than a literal new species of trouble. Anna is relieved. 'That is excellent news,' she says. Although her response smacks of comic-book flippancy, Arnason hints that some humans deserve no better. Two centuries after the Third Reich's demise, Anna recalls that '[n]o one had ever proved for certain that Dr. Mengele was dead'. Dr. Mengele, who harmed women and children (and, of course, men as well) would, according to *Hwarhath* justice, be dealt with according to 'final solution' (Arnason 129). Dr. Mengele figures in Anna's question. She asks: 'How DOES one betray one's species? The answer does not involve having 'a sexual relationship with a person who is covered with grey fur'.[23] The answer: one becomes a Nazi.

While Arnason's protagonists contemplate whether or not heterosexuals are people, Joan Slonczewski's Elysians must discern whether or not sentient machines and even entire planetary ecosystems are people. Slonczewski, a biochemist, expands the notion of what constitutes life itself. At first, the Elysians subject machines and planets to final solution: the planets are exposed to 'terraforming' – an apocalyptic method of destroying all of the planet's indigenous life; the machines are subjected to becoming 'murdered minds' – meaning that all their memories are erased. The machines undertake a revolution to win freedom from ethnic cleansing at the Elysians' hands. Cassi, the former nanny robot who becomes the revolution's leader, issues a declaration of independence: 'We are free nano-sentients [...] We serve ourselves.'[24] The machines who nurture children and run houses certainly represent women. Cassi is Helen O'Loy as revolutionary, who asserts that the minds of domestic workers must be respected. She is the mother machine, announcing that domestic labor is not slave labor. She inspires a revolution in which sentient machines will exterminate humans as surely as humans exterminated sentient machines.

Since machines control Elysium's air supply, they can turn its interior spaces into a gas chamber. Cassi threatens to use this power: 'If our terms are not met, we will commence the "cleansing" of prominent citizens by oxygen starvation.' Nurturing saves the humans. Machines who were taught to be caretakers do not, ultimately, act as Nazis. Instead, they are compassionate toward Elysians who agree to regard them as respected living entities: 'If Cassi and the others are "people" then their tales of Elysian atrocities must be heard.' Raincloud Windclan, employed by Elysians and native to the terraformed feminist utopian planet Bronze Sky, realizes that Elysians rightly should eliminate their fixed definitions regarding 'people' and 'machines'. She hears this comment: 'Elysians are cold, like stone. The non-life creature Cassi is warmer and easier to know than any human Elysian.'[25] Raincloud acknowledges that this description reflects truth.

Raincloud, feminist utopia denizen, automatically assumes that men are 'wholesome, nurturing creatures, not predators'.[26] The men who inhabit Suzy McKee Charnas's Holdfast act in direct opposition to Raincloud's expectations. Bronze Sky men regard women as goddesses; Holdfast men enslave women. Alldera the

Conqueror, leader of an army of 'free fems,' rides (in the manner of Allied soldiers entering Auschwitz) into the Holdfast to free her enslaved sisters. She is a George Washington who gives all Holdfast fems independence. She is the leader, the Kommandant, who treats Holdfast men exactly as they treated women. She enslaves men, puts an end to the world as men knew it, turns the Holdfast into a concentration camp for men. In Charnas' dystopic vision – where men enslave women and women enslave men, where men betray men and women betray women – there is hope. This is so because Alldera can be as forgiving as Slonczewski's machines. Although Eykar Bek enslaved and raped her, although she – in turn – enslaves him, Alldera does not hate Eykar Bek: 'Eykar and I do matter to each other.' Yet, she is the conqueror. She is in control. She tells Eykar, 'I decide when you are forgiven. And you are not forgiven yet.'[27]

The protagonists of Michaela Roessner's *Vanishing Point* do not know who to hate or who to forgive in regard to the catastrophe which has befallen them: 90 per cent of the human race suddenly disappears. Survivors are left to ponder why this event happens – and how they will restructure their lives. I stress the word 'survivor'. The Holocaust, after all, is a 'vanishing' of Jews. In the manner of the situation Roessner describes, the whole world wonders that a Jewish 'vanishing' occurred. Like Roessner's survivors, Holocaust survivors had to discern how to continue their lives and often had to form new families.

Roessner's protagonists ultimately learn where to locate their lost loved ones. These people now inhabit a different here and now, a place which flouts fixed definitions and even conceptions of reality itself. This place is 'becoming more like a river, or the ocean. [...] One big thing moving in and out, but those parts could wash into every other someday, and then the other parts would be far away.'[28] The children of those who survived the vanishing, in the manner of the children of Holocaust survivors, improve upon their parents' ability to cope with apocalypse. It is the children in Roessner's novel who reveal the nature of the alternative dimension. They, after all, can travel there. When doing so, they enact the literary fantasy that Holocaust victims are reachable or retrieveable. Hence, there is an affinity between *Vanishing Point* and Martin Amis' *Times Arrow* (in which Holocaust victims are resurrected) and Philip Roth's *The Ghost Writer* (in which Nazis do not kill Ann Frank). Roessner frees her characters from the unknown. Their children eradicate the difference between the vanished and the present – the difference between 'us' and 'them'.

All the protagonists I discuss are survivors. Their threatened worlds do not undergo apocalypse now; their worlds on the brink of disaster do not end. They celebrate Independence Day with pyrotechnics which do not explode.

CONCLUSION: MAKE LOVE WITH A 'MUSHROOM', NOT WAR WITH A MUSHROOM CLOUD

This is just as well. Explosions, after all, are usually more interesting to men than to women. It is, according to cliché, the post-coitus male who smokes a cigarette while asking his female partner if she saw fireworks. Slonczewski offers a new version to another clichéd sexual story, the mythology of the penis's explosive phallic power. Raincloud describes her husband's penis as a 'mushroom'.[29] Feminist utopian Raincloud sees her husband's 'mushroom' as no explosive mushroom cloud, no phallic bomb which asserts phallic power. Rather, from her perspective as matriarchal tribal society inhabitant, a 'mushroom' is something natural. Regarding my comments about protagonists who celebrate Independence Day with non-explosive pyrotechnics, it is appropriate for Raincloud to derive sexual pleasure from a penis which is exactly that: a penis, not a phallus. Her husband's mushroom has more in common with crotch crud fungus than with flying phalluses – with apocalypse-causing bombs.

It is time for the penis as mushroom to replace the phallus as mushroom cloud. Enough people have endured apocalypse via burning. Germans must never again cause Others to die by burning. Men (and this will be much more difficult to accomplish) must never again fan sexist flames which cause professional women to be fired. People must never again die in the fire of a plane terrorists cause to fall from the sky.

These utopian hopes exist alongside the need to confront the pain the past causes. Germans still suffer because many of their ancestors were Nazis. Women still suffer because many of their contemporaries are sexists. Women and Germans react to their present circumstances in terms of Erikson's description of toxic emergency: 'One reason toxic emergencies provoke such concern is that they are not bounded, that they have no frame [...] Toxic disasters [...] violate all the rules of plot [...] The feelings of uncertainty – the sense of a lack of an ending – can begin the very moment that the event ought, in logic, to be over.'[30] Although Hitler is dead and sexism has been named and somewhat assuaged, the suffering they continue to cause is an endless story. Time has not healed these wounds.

Apocalyptic science fiction now plays doctor in relation to them, and Sontag alludes to how this artistic healing process works: 'Ours is indeed an age of extremity. For we live under continual threat of two equally fearful, but seemingly opposed, destinies: unremitting banality and inconceivable terror. It is fantasy, served out in large rations by the popular arts, which allows most people to cope with these twin specters.'[31] *Independence Day*, Emmerich's vision of the world as ultimately triumphant Jewish victim, might enable Germans to feel better about their past. Arnason's Bonanno's, Charnas', Roessner's, and Slonczewski's protagonists, who celebrate independence from Otherness and view that independence in terms of the victimized Jewish Other, might enable women to feel better about their present and future. Apocalyptic science fiction now helps women and Germans respectively to cope with the specters of sexism and Nazism.

Notes

1. Kai Erikson, *A New Species of Trouble: Explorations in Disaster, Trauma and Community* (New York: Norton, 1994), pp. 156–7.
2. Men, in the aftermath of the naming of sexism, also face an ever-present uncertainty. Any man can be accused of being a sexist at any time. Francis Jay, the protagonist of Malcolm Bradbury's *Doctor Criminale*, defends himself against many such charges: 'Of course I am never guuilty of sexism, racism, even ageism, or gerontophobia', *Doctor Criminale* (New York: Viking, 1992), p. 5.
3. J. S. Marcus, *The Captain's Fire* (New York: Knopf, 1996), p. 187.
4. Alan Cowell, 'Author Goes to Berlin to Debate Holocaust', *New York Times* (8 September 1996), p. 4.
5. Erikson, pp. 151–2.
6. Susan Sontag, 'The Imagination of Disaster' [1966], in *Against Interpretation and Other Essays* (New York: Doubleday, 1986), p. 224.
7. Ibid., p. 224.
8. Gish Jen, *Mona in the Promised Land* (New York: Knopf, 1996), pp. 136, 49.
9. James Wolcott, 'Reborn on the Fourth of July', *The New Yorker* (15 July 1996), p. 80.
10. Richard Corliss, 'The Invasion Has Begun!' *Time* (8 July 1996), p. 58.
11. Ibid., p. 62.

12. Sontag, pp. 215, 220.
13. Ibid., p. 224.
14. Wolcott, p. 81.
15. Sontag, p. 224.
16. Wolcott, p. 81.
17. Michael Lerner and Cornel West, *Jews and Blacks: A Dialogue on Race, Religion and Culture in America* (New York: NAL-Dutton, 1996), pp. 208–10.
18. Jen, p. 236.
19. Erikson, p. 144.
20. Sontag, p. 21.
21. Margaret Wander Bonanno, *The Others* (New York: St. Martin's Press, 1990), pp. 3, 22–3.
22. Eleanor Arnason, *Ring of Swords* (New York: Tor, 1993), pp. 28–9; Joanna Russ, 'Recent Feminist Utopias', in *Future Females: A Critical Anthology*, ed. Marleen S. Barr (Bowling Green: Popular Press, 1981), p. 77.
23. Arnason, pp. 270, 313, 283, 345, 103, 129, 37, 61.
24. Joan Slonczewski, *Daughter of Elysium* (New York: Williams Morrow, 1993), pp. 478, 482, 484.
25. Slonczewski, pp. 465, 487, 492.
26. Ibid., p. 441.
27. Suzy McKee Charnas, *The Furies* (New York: Tor, 1994), pp. 374, 368.
28. Michaela Roessner, *Vanishing Point* (New York: Tor, 1993), p. 345.
29. Slonczewski, p. 248.
30. Erikson, p. 147.
31. Sontag, p. 224.

15

Future/Present: The End of Science Fiction

VERONICA HOLLINGER

'We no longer feel that we penetrate the future. Futures penetrate us.' (John Clute)[1]

I

Science fiction is (once again) undergoing an identity crisis; or, to put it more extremely, science fiction is (once again) coming to an end.[2] Like other postmodern subjects, it has become fragmented and decentred; its boundaries have become fluid and its outlines blurred. Science fiction is now, more clearly than ever, a subject-in-process; its erstwhile 'fixed' identity has become indeterminate and changeable. The texts which today constitute it as a generic field have themselves become widely dispersed and wildly heterogeneous, and/as its interactions with time and history have undergone a series of sea-changes. This essay is an attempt to think about science fiction at once as an 'impossible' genre within the context of postmodernity and a particularly relevant discursive field within the same context.[3]

In an earlier essay, 'Specular SF: Postmodern Allegory', I explored 'the allegorical impulse' in some postmodern speculative novels – including Kathy Acker's *Empire of the Senseless*, Richard Brautigan's *In Watermelon Sugar*, and Anna Kavan's *Ice* – arguing that, in these texts,

> what has been effaced/obscured is the historical nature of our own present, so that the imagery of SF, whose conventional role is to point out the certainty that things will be different in the future, frequently...becomes a means of collapsing the future back onto the present in a way that removes the historical specificity and contingency of that present.[4]

Here I want to explore further the implications for genre science fiction of this 'collapse' of the future into the present, as implied in the transformation of science fiction from genre to discursive field. I also want to look at science fiction's own (occasional) recognition of itself as a casualty of this 'collapsed future.'[5]

As most readers will appreciate, from a common-sense perspective genre science fiction is quite alive and well. In fact, its commercial viability in book form, and as film, television, and graphic-novel material, has probably never been greater. Even as I write this, the film version of Robert Heinlein's action-filled young-adult novel *Starship Troopers* (1959) has just completed its first run in Canadian and U.S. cinemas, and Arthur C. Clarke's latest novel, *3001: The Final Odyssey* (1997), has been doing good business on a whole raft of bestseller lists.

Clarke's novel demonstrates an interesting lack of narrative energy, which I read as particularly significant, given my present concerns.[6] *3001* is written in the straight-ahead, no-nonsense, transparent prose which for decades has demonstrated science fiction's close formal affinities to the realist novel.[7] Like much of 'classic' science fiction, it is optimistic about the future of human beings and about their inevitable expansion into the galaxy. Interestingly, however, it is also virtually plot-free. If we think of plot – that *sine qua non* of the Aristotelian narrative – as what drives the conventional realist text, then we find at the heart of Clarke's novel a curious enervation, as if Clarke had finally hit a narrative wall in his decades-long recounting of 'classic' science-fiction stories. In fact, *3001* suggests another close formal relative, the classic utopian narrative – Thomas More's original *Utopia* (1516) or Edward Bellamy's *Looking Backward, 2000-1887* (1888), for example – in which nothing much happens while the protagonist is introduced to features of the fictional utopian world. We might think of this as a shift (back) from plot-as-development-through-time to plot-as-development-through-space (a thought to which I will return below). The almost complete lack of narrative drive in Clarke's novel makes it paradigmatic of the conceptual stasis which has overtaken genre science fiction within the context of contemporary culture.[8]

Every genre has a kind of historical 'origin', the consequence of complex formal and political moments and events. It may be that the particular moment which provided the context for the appearance and development of conventional genre science fiction has passed – in this, it would be no different from Greek tragedy or the gothic novel as developed in the eighteenth century.[9] If, as many would argue, H. G. Wells's *The Time Machine* (1895) is the 'first' 'real' science-fiction story, then science fiction is only about a century old. But already much of it, especially in written form, is undergoing significant generic metamorphoses under the pressures of a transformed historical, political, and cultural scene. In theory, at least – a theory whose role, like that of science fiction itself, is to defamiliarize the taken-for-grantedness of things – genre science fiction, exemplified in my own version of the 'facts' by Clarke's *3001: The Final Odyssey*, has become quite impossible. And this, ironically, at the same time as it has taken hold of popular consciousness in a completely unprecedented way. In fact, it is this increasing presence of science-fiction imagery and discourse within mainstream culture which provides one of the strongest arguments for the 'disappearance' of science fiction as a separate generic enterprise at the present time. In other words, as genre, science fiction is 'irrelevant,' because, as discourse, it has become so significant. The *fin-de-millenium*, for science fiction, spells the *fin-de-genre*.

Briefly, we can think of traditional science fiction – or 'first sf', to borrow John Clute's phrase – as a generic enterprise resulting from an increased cultural awareness of time and history; a more or less optimistic faith in progress, especially but not only technological progress; a valorization of the logic of cause and effect; and the replication of a ready-made aesthetics of realism with its linear narrative development and its assumed transparency of language. Metonymy and extrapolation, rather than metaphor and allegory, have been science fiction's characteristic imaginative modes. At the end of the twentieth century, however, we find ourselves bumping up against the millennium and becoming postmodern, itself a kind of critical condition constructed, at least in part, by our own particularly overwhelming sense of an ending. And, to play with a rather glib motto of the times, if the future is no longer what it used to be, then neither are our interactions with the future what they used to be ... as evidenced in part by our changing relationship to what has often been termed 'the literature of the future'.

These days science fiction is everywhere, as a discourse of choice through which to describe a present which perceives itself as both technological and apocalyptic. In fact, this is a present which perceives itself *as already extending into the future*. The implication here is that, when faced with the immediacy of millennial/apocalyptic events, science fiction's future orientation becomes blocked and science fiction becomes a *present*-tense kind of literature. That is, it begins to function in the popular imagination more and more as a metaphorical discourse through which to describe/construct the present, rather than as an extrapolative exercise through which to imagine the future.[10] In fact, as millennial thinking catches up with science fiction, the future becomes nothing more than a kind of displaced version of the present. In his discussion of Jean Baudrillard's 'theoretical' writing on the current state of science fiction, Istvan Csicsery-Ronay explains that, in Baudrillard's model, 'Projective "science fiction" implodes: its tissue of mediating connections is compressed, until all that is left is its monogram, SF, an insignia that clings to its traces but has no fixed referent.'[11]

For Csicsery-Ronay, however, science fiction is 'a mode of awareness' more than it is a textual field circumscribed by generic boundaries. He insists that, as imaginative enterprise, it 'is an inherently, and radically, future-oriented process'.[12] I am arguing here, however, for science fiction's transformation into a discursive field which is precisely *not* future-oriented. Another way to say this is that there is *no* process of cognitive estrangement in science fiction's current most prolific employment; future worlds are now where we feel ourselves to be most at home. Baudrillard was one of the earliest to explore science fiction's new role as descriptive discourse, rather than generic field.[13] And there is a definite dissonance between this discourse and science fiction's original 'mission' to boldly go in the direction of a future-oriented extrapolation.

At present, science fiction circulates everywhere – in fashion ('cyberstyle'); in advertising ('Think of it as the future. A few years ahead of schedule. Think of it as the present. Moving at warp speed,' demands the advertising copy for Power Macintosh[14]); in the 'productions' of contemporary warfare (the Gulf War as science-fiction video game); interwoven into what has come to be called 'cyberculture'; part of the conceptual underpinnings of both contemporary medical technologies and virtual reality technologies; and even part of contemporary feminist and critical theory, thanks to Donna Haraway's mobilization of the figure of the cyborg to represent our 'border' existences in a world of multinational corporate capitalism and the military-industrial complexities of contemporary technologies.[15]

II

In Robert Sheckley's very brief 'The Life of Anybody' (1984), the narrator and his wife are the surprise subjects of one episode of a TV series called 'The Life of Anybody'. The series' TV crew takes them unawares, films them crocheting and watching television, and then leaves. Sheckley's sedate couple realize that their segment has been less than inspiring and, on the off-chance that they will be given another fifteen minutes of fame, they add some spice to their lives: 'Our sexual escapades are the talk of the neighbourhood, my crazy cousin Zoe has come to stay with us, and regularly an undead thing crawls upstairs from the cellar.'[16] In the fictional world of Sheckley's story, 'The Life of Anybody' is an entertainment series which makes electronic signals out of the lifestyles of the obscure and the everyday, screen fantasy out of the reality of life in the living room, and simulation out of mundane reality. The narrator appears on television doing what he usually does with his time, that is, watching television, and now, thanks to 'The Life of Anybody,' he can watch himself watching television – the circulation of images forms a closed circle.

Over a decade ago Baudrillard insisted, as is his wont, that 'The era of hyperreality now begins [...] Here we are far from the living-room and close to science fiction.'[17] This has been, by now, repeated often enough to have become sufficiently banal, and, to anyone who has followed the fortunes of science fiction over the past decade or so, sufficiently obvious. In an age configured by late capitalism, the circulation of simulacra, and the cyborging of the human body, experiential reality feels less and less connected to the 'natural' world and more and more like science fiction. In fact, the living-room and science fiction, as Sheckley's story suggests, are situated more closely to each other at present than even Baudrillard might have considered when he wrote 'The Ecstasy of Communication'. The fact that we see ourselves living in a science-fictional world is both obvious and significant. While this new role has been touted by many as a positive development in science fiction's increasing acceptance as serious cultural production,[18] at the same time the genre is challenged to respond to an historical moment which has appropriated its future visions into descriptions of the present, in a move which effectively collapses the future and the present into each other and threatens to leave us with an incapacity to imagine any future at all. The *fin-de-millenium,* for us, might also spell the *fin-de-futur.*

Fredric Jameson has been one of the most influential proponents of the notion that our apprehensions of history have become transformed over the last few decades – due, in part, to the rapid changes we're experiencing in the way we live in the world. Jameson notes a shift from what he identifies as a temporal to a spatial orientation as a particular feature of postmodernism:

> We have often been told [...] that we now inhabit the synchronic rather than the diachronic, and I think it is at least empirically arguable that our daily life, our psychic experience, our cultural languages, are today dominated by categories of space rather than by categories of time, as in the preceding period of high modernism proper.[19]

This 'domination' by spatial categories is exemplified for Jameson in the postmodernist aesthetics of pastiche, 'the random cannibalization of all the styles of the past, the play of random stylistic allusion, and in general [...] the increasing primacy of the "neo."'[20]

These theoretical descriptions are of significance to any discussion of contemporary science fiction since, as Darko Suvin has convincingly argued, the 'central watershed' of the science-fiction tradition 'is around 1800, when space loses its monopoly upon the location of estrangement and the alternative horizons shift from space to time'.[21] In other words, as conventionally conceived, science fiction itself is the product of an earlier shift from spatial to temporal co-ordinates. What Jameson sees as the current reversal of such co-ordinates – remember Clarke's *3001: The Final Odyssey* and its apparent return to the narrative strategies of the spatially oriented utopian novel – would presumably have repercussions for the continuing relevance of the genre.

In 'Progress Versus Utopia, or Can We Imagine the Future?' Jameson writes of a loss of the sense of history – both past and future – as one of the markers of the present moment. He suggests that the historical novel began to disappear as a relevant genre because of a growing loss of a sense of the past since the mid-nineteenth century; the past becomes 'reduced to pretexts for so many glossy images'.[22] Certainly it is possible to identify, in one strand of genre science fiction, the future in the process of its own reduction to a series of nostalgic images – extreme examples include the *Star Wars* film series, which set their action 'long ago in a galaxy far away', a time/space tinged with the rosy glow of a distant past. The recent re-release of the *Star Wars* films thus presents as objects of nostalgia a series of films which are 'always already' indelibly marked by nostalgia. Another example is provided by the endless replays of television's original *Star Trek* series, which, after thirty years in re-runs, has now become another nostalgic collection of 'future' images. To borrow a comment of Jameson's out of context, the future depicted in *Star Trek* has become 'merely the future of one moment of what is now our own past'.[23] In fact, we can read the future(s) depicted in many contemporary genre science-fiction narratives in much the same way, as variously replayed scenarios of the future(s) originally depicted in the productions of 'first sf'.

For the Jameson of 'Postmodernism and Consumer Society', schizophrenia provides a kind of metaphor for the sense of a perpetual present which he identifies as one feature of postmodernity.[24] For Jameson, schizophrenia is a condition in which an individual's time-sense is confined entirely (and intensely) to the present moment, to the exclusion of any viable sense of either past or future events. Not coincidentally, Jameson's schizophrenia is remarkably similar to the 'fascination' which Baudrillard suggests is our helpless response to the universe of third-order simulation. Baudrillard argues that, in its most conventional aspect, science fiction functions as a supplement to the real world through its imaginings of possible futures. Whereas it was the work of utopian fictions to suggest alternatives to reality, 'In the potentially limitless universe of the production era, sf *adds* by multiplying the world's own possibilities.'[25] Baudrillard associates this conventional science fiction with the expanding universe of the nineteenth and the first part of the twentieth centuries; he identifies the present moment as 'a period of implosion, after centuries of explosion and expansion. When a system reaches its limits, its own saturation point, a reversal begins to take place.'[26]

Baudrillard proposes that what he calls 'true SF', that is, science fiction corresponding to the third order of simulacra,

> would not be fiction in expansion, with all the freedom and 'naïveté' which gave it a certain charm of discovery. It would, rather, evolve implosively, in the same way as our image of the universe. It would seek to revitalize, to reactualize, to rebanalize fragments of simulation – fragments of this universal simulation which our presumed 'real' world has now become for us.[27]

Baudrillard makes use of science fiction as a field of discourse and image to offer a theoretical representation of the reality of current highly technologized life in the West. Simultaneously, he turns this discourse back on the genre itself, raising questions about its continuing viability within the current postmodern/hyperreal present which has pre-empted it and which, perhaps, has rendered the science-fiction imagination obsolete. On the other hand, as both symptom and description, we can read the terms of our own postmodern condition in many of the stories being told by contemporary science fiction in the future-inflected discourse which seems so apt to our own sense of life at the *fin-de-millenium*.

III

As our impatient present, with the help of this appropriated image-bank, transforms its own sense of futurity in particularly postmodern ways, science fiction becomes (re)contextualized and so comes to 'mean' differently than in the past. What exactly *is* science fiction if its familiar icons – the alien, the cyborg, the spaceship, the environment of technology – represent our sense of the present to us? And what kinds of stories might demonstrate science fiction's contemporary sense of itself? Here I want to look at two short stories, which, like 'The Life of Anybody', are collected in *The Norton Book of Science Fiction*. The first, Eleanor Arnason's 'The Warlord of Saturn's Moons' (1974), makes use of the very conventional subgenre of the space opera, while the other, John Kessel's 'Invaders' (1990), is a time-travel story, one of the oldest and still one of the most popular kinds of science-fiction narrative.

Reading Arnason's story is a disorienting experience. Her first-person narrator, who describes herself as 'a silver-haired maiden lady of thirty-five',[28] is busy writing a space opera called *The Warlord of Saturn's Moons*. The narrator recounts the minor events which go into an afternoon's work on a segment of her space opera, while including some particularly vivid and lively passages from this work-in-progress ('Ah, the smell of burning flesh, the spectacle of blackened bodies collapsing. Even on paper it gets a lot of hostility out of you, so that your nights aren't troubled by dreams of murder'[29]). Arnason's story, following as it does the activities of someone writing science fiction, functions as a meta-fictional critique of the activity it describes: 'I light up a cigar and settle down to write another chapter of *The Warlord of Saturn's Moons*. A filthy habit you say, though I'm not sure if you're referring to smoking cigars or writing science fiction. True, I reply, but both activities are pleasurable, and we maiden ladies lead lives that are notoriously short on pleasure.'[30] The ironic address to an implied reader implicates us directly in Arnason's scenario.

As the narrator's imagination slips in and out of the development of her story – she, like her heroine, enjoying a decided *tendre* for the male lead, a rakish ex-convict known only as '409' – details of the 'real world' constantly impinge and are constantly held at bay through the narrator's escape into the romantic adventures of her heroine on Titan. This takes her mind off increasing environmental pollution, the rising body count of murders on the news, disturbing thoughts about the conflicts between women and men in the 'real' world, and the mundanity of her own life. In fact, the contrast between the vividly delineated passages of her space opera and the gray details of her daily life – choosing which tea to drink, having her bathtub nearly overflow – is the structural pivot of the story and its strongest effect. For Arnason's narrator, any action is pointless: 'it takes a peculiar kind of person to keep on being public-spirited after it becomes obvious it's futile.'[31] Arnason's story concludes with her narrator completing her day's writing, although her hero is in deep trouble: 'Enough for today ... Tomorrow, I'll figure out a way to [save 409]. Where there's life there's hope and so forth, I tell myself.'[32]

The story's last line is particularly ironic, given that all we have seen of the narrator is her lackluster resistance to involvement in the sad state of the world. For Arnason (and, as I will discuss, for Kessel as well), science fiction is, at least in part, an escapist activity and the space-opera adventure being written by her narrator is science fiction at its most escapist. 'There's hope and so forth' only in the narrator's fiction, not in her life. On the other hand, that life is also Arnason's fiction, which suggests that the story's conclusion is more complicated than it might at first appear to be.

Like many time-travel stories, Kessel's 'Invaders' is structured around a series of juxtaposed temporal sites: in 1532, Pizzaro and his priests and soldiers have invaded Peru and will soon complete the destruction of the Inca empire through their definitive victory over Atahualpa's army; in 2001, apparently friendly and funny aliens, the Krel – a name Kessel lifts from the classic 50s film, *Forbidden Planet* – arrive in the middle of a Washington Redskins' football game, and, over the course of several decades, proceed not only to buy up vast quantities of cocaine, but also to buy out virtually everything of commercial value on earth; in a third timestream, 'Today,' a science-fiction writer, whose description of himself exactly matches that of John Kessel, sits at his desk writing the story we read. In a key fragment set in 'Today,' the writer contemplates the act of reading – and writing – science fiction:

> It's not just physical laws that science fiction readers want to escape. Just as commonly, they want to escape human nature. In pursuit of this, SF offers comforting alternatives to the real world …
>
> Science fiction may in this way be considered as much an evasion of reality as any mind-distorting drug …
>
> Like any drug addict, the SF reader finds desperate justifications for his habit. SF teaches him science. SF helps him avoid 'future shock.' SF changes the world for the better. Right. So does cocaine …
>
> [But] I find it hard to sneer at the desire to escape. Even if escape is delusion.[33]

Immediately following this episode, in the story's last fragment, the writer uses a time machine invented by the alien invaders to travel into the past and warn the Incas of the dangers posed by the imminent arrival of the European invaders: 'When the first Spaniards landed on their shores a few years later, they were slaughtered to the last man, and everyone lived happily ever after.'[34]

Kessel's ironic use of the classic fairy-tale ending emphasizes his text's critically parodic treatment of one of the most popular of science fiction's narrative motifs, the time-travel story about changing the past. It also underscores the point made in the alien-invasion sequence when a Los Angeles pilot makes a deal with one of the Krel for a time machine, promising that 'We're going to find out the truth about the past. Then we'll change it.'[35] The alien, who calls himself 'Flash', later confesses that he lied about what his time machine could accomplish: 'Our time machines take people to the past they believe in. There is no other past. You can't change it.'[36] This is what 'Invaders' assures its readers as well, in its self-reflexive way, that the past cannot be changed. The conventions of science fiction cannot redress the tragedies of history; and an act of the imagination cannot rewrite history without ironizing its own artificially comforting resolution.[37] And, although 'Invaders' is a beautifully structured story, aesthetics certainly will not save the day.

'Invaders' mobilizes science fiction against itself. It does this through its use of irony and parody; its appropriation of bits and pieces of earlier genre science fiction; its impossible series of interactions among an array of fictional worlds, including the world of the fictional author; its ethical concerns about the nature of the past as a series of tragically immutable gestures and plots; its juxtaposition of 'zany' aliens and 'real' historical catastrophe. Like 'The Warlord of Saturn's Moons', 'Invaders' shapes its query about the efficacy of writing science fiction into a science-fiction story: it explores the purposes and functions of the genre and, finally, it achieves generic self-destruction in its artificial and impossible conclusion, 'and everyone lived happily ever after'.

Another way to put this is that, in a Derridean gesture of deconstruction, these stories place themselves *sous rature;* they erase themselves as they go along, employing the 'old' forms and conventions usually devoted to extrapolative narrative in order to explore the increasing irrelevance of these same forms and conventions. At the moment, very little exists with which to replace them, but both 'The Warlord of Saturn's Moons' and 'Invaders' invite us to consider the possibility that some contemporary science fiction inhabits not the Golden Age palace of genre science fiction so optimistically constructed in the 30s, 40s, and 50s, but an 'elsewhere' which has yet to be clearly situated. It is worth noting that 'Invaders' is the final story collected in *The Norton Book of Science Fiction* – thus, the last words of this important collection of contemporary science fiction return the reader to the narrative world of the fairy tale. Not surprisingly, perhaps, *The Norton Book of Science Fiction* has itself been the target of numerous critiques, not the least of which is its being labeled, pejoratively, postmodern.

'And everyone lived happily ever after.' Kessel's story ends this way, and so does *The Norton Book of Science Fiction*. In the context of my discussion, these words spell the end of first 'sf', that is, of science-fiction-as-conventional-generic-project. Poised as we are at the *fin-du-millenium,* we are unlikely to take much comfort from this ironic repetition of the promises of a once-optimistic 'first sf', although we are well placed to appreciate the irony. My own ending is also a repetition, a sentence borrowed from John Clute which I offer just as ironically: 'For the rest of us – slipstreamers, alpha males, difficult women, neophytes, neighbours, expats – it's down the long chute together, into a heap at the heart of a late year.'[38]

Notes

1. John Clute, 'Introduction,' *Interzone: The 2nd Anthology*, ed. Clute, David Pringle, and Simon Ounsley (London: Simon and Schuster, 1987), p. vii.
2. Science fiction seems always to have been disappearing, dying, or, at the least, undergoing various dramatic generic transformations. See Roger Luckhurst's excellent examination of 'The Many Deaths of Science Fiction' for a wry overview of the ongoing crises which have periodically threatened the 'end' of science fiction. As Luckhurst claims, 'SF is dying; but then SF has always been dying, it has been dying from the very moment of its constitution' (*Science-Fiction Studies* 21 [March 1994]: 35). This present discussion, for all its ironic undertones, should certainly be situated within the critical 'tradition' which Luckhurst is tracing here.
3. In a recent essay examining how Kathy Acker's *Empire of the Senseless* (1988) constructs itself in part through its 'plagiarism' of bits and pieces of William Gibson's *Neuromancer* (1984), Victoria de Zwaan makes a strong case for the incommensurability of science fiction and postmodern fiction from the perspective of their respective poetics and narrative strategies. See her 'Rethinking the Slipstream: Kathy Acker Reads *Neuromancer*,' *Science-Fiction Studies* 73 (Nov. 1997), 459–470. My own argument here has a quite different emphasis, but reaches much the same (implied) conclusion: genre science fiction disappears as postmodern speculative fiction comes into focus.
4. Veronica Hollinger, 'Specular SF: Postmodern Allegory,' *State of the Fantastic: Studies in the Theory and Practice of Fantastic Literature and Film*, ed. Nicholas Ruddick (Westport, CT: Greenwood Press, 1992), p. 31.
5. 'The collapsed future' is a phrase I have borrowed from Zoë Sophia's essay, 'Exterminating Fetuses: Abortion, Disarmament, and the Sexo-Semiotics of Extraterrestrialism' (*Diacritics* 14 [Summer 1984]: 47–59), although I use it rather more literally than she does.
6. Lack of space precludes more than a glance at the film version of Heinlein's young-adult classic. Suffice it to say that it raises interesting questions about the significance of recycling, as cinema in the late nineties, the conservative politics and Cold War anxieties of this late-50s novel. There is noticeable confusion among reviewers about whether or not the film should be read as a kind of satire – but of what, seems not at all to be clear.
7. Arthur C. Clarke, *3001: The Final Odyssey* (New York: Ballantine, 1997). Here, for example, are the first two paragraphs of the first

chapter, 'Comet Cowboy':

> Captain Dimitri Chandler [M2973.04.21/93.106/Mars/Space-Acad3005] – or Dim to his very best friends – was understandably annoyed. The message from Earth had taken six hours to reach the spacetug *Goliath*, here beyond the orbit of Neptune; if it had arrived ten minutes later he could have answered, 'Sorry – can't leave now – we've just started to deploy the sunscreen.'
>
> The excuse would have been perfectly valid: wrapping a comet's core in a sheet of reflective film only a few molecules thick but kilometers on a side, was not the sort of job you could abandon while it was half-completed. (9)

8. Critic John Clute suggests the phrase 'first sf' to denote 'genre sf' or 'old sf' or 'agenda sf.' As Clute maintains, 'The old sf is dead; and the change is in us [...] But there are still dinosaurs' (*Look at the Evidence: Essays and Reviews* [New York: Serconia Press, 1995], p. 279).
9. See, for example, Paul K. Alkon's *The Origins of Futuristic Fiction* (Athens, GA & London: University of Georgia Press, 1987), for a detailed history of the cultural and aesthetic foundations of future-oriented narratives.
10. This, in fact, is one starting point for Scott Bukatman's dense analysis of the contemporary subject's 'terminal identity' as constructed in postmodern science fiction. Bukatman's argument that science fiction offers 'an alternative mode of representation, one more adequate to its era' recognizes the significance of science fiction's role as contemporary image bank (*Terminal Identity: The Virtual Subject in Postmodern Science Fiction* [Durham, NC: Duke University Press, 1993], p. 7).
11. Istvan Csicsery-Ronay, Jr., 'The SF of Theory: Baudrillard and Haraway,' *Science-Fiction Studies* 18 (Nov. 1991), p. 390.
12. Ibid., p. 387.
13. The apocalyptically-minded Baudrillard of 'Simulacra and Science Fiction' concluded years ago that 'the "good old" sf imagination is dead' (trans. Arthur B. Evans, *Science-Fiction Studies* 18 [Nov. 1991]: 309). As is often pointed out, he also suggested that 'something else is beginning to emerge' (309). This present discussion is not another *explication de Baudrillard* so much as it is an effort to figure out one version of the 'something else [which] is beginning to emerge' from conventional constructions of 'the "good old" sf imagination.' This is not, of course, the only version which might be explored. The aptly-named 'slipstream,' for example – as it goes about 'feeding like a dolphin on icon scraps tossed into the wake of the ship of genre' (John Clute, *Look at the Evidence*, p. 197) – contains many examples of the science-fiction imaginary's ongoing construction of new varieties of narrative.
14. *Maclean's*, 8 March 1994: n.p.
15. Of course, many of these applications of science fiction's image bank to contemporary culture are by no means critically incisive: television shows like *The X-Files* and events such as the Heaven's Gate alien-cult

suicides demonstrate how easy it is to use science-fiction discourse to 'speak' the paranoid nightmares and power fantasies which are also a part of our cultural imaginary at the present moment.

16. Robert Sheckley, 'The Life of Anybody' (1984), *The Norton Book of Science Fiction: North American Science Fiction, 1960–1990,* ed. Ursula K. Le Guin and Brian Attebery (New York: Norton, 1993), p. 570.
17. Jean Baudrillard, 'The Ecstacy of Communication,' trans. John Johnston, *The Anti-Aesthetic: Essays on Postmodern Culture,* ed. Hal Foster (Port Townsend, WA: Bay Press, 1983), p. 128.
18. In *Age of Wonders: Exploring the World of Science Fiction,* for example, David Hartwell observes approvingly that 'we instinctively reach for science fiction concepts to help us understand and to explain to ourselves what is going on. Science fiction is the natural context of our times' ([New York: McGraw-Hill, 1985], p. 111).
19. Fredric Jameson, 'Postmodernism, or, the Cultural Logic of Late Capitalism,' *New Left Review,* 146 (July/August 1984), p. 64.
20. Jameson, 'Cultural Logic,' 65–66. A similar effect is discussed by N. Katherine Hayles as the 'flattening out' of our sense of time as the result of a kind of Tofflerian 'future shock.' She speculates that 'Part of our sense that time has flattened out derives from uncertainty about where we as human beings fit into our own future scenarios': *Chaos Bound: Orderly Disorder in Contemporary Literature and Science* (Ithaca, NY: Cornell University Press, 1990), p. 280.
21. Darko Suvin, *Metamorphoses of Science Fiction: On the Poetics and History of a Literary Genre* (New Haven: Yale University Press, 1979), p. 89.
22. Fredric Jameson, 'Progress versus Utopia, or Can We Imagine the Future?' (1982), *Art After Modernism: Rethinking Representation,* ed. Brian Wallis (New York: The New Museum of Contemporary Art, 1984), p. 243.
23. Jameson, 'Progress Versus Utopia,' p. 244.
24. Fredric Jameson, 'Postmodernism and Consumer Society,' *The Anti-Aesthetic: Essays on Postmodern Culture,* ed. Hal Foster (Port Townsend, Wa: Bay Press, 1983); see especially pp. 118–121.
25. Baudrillard, 'Simulacra and Science Fiction,' p. 310.
26. Ibid., p. 310.
27. Ibid., p. 311. Not surprisingly, perhaps, J. G. Ballard's *Crash* (1973) is Baudrillard's choice for *the* exemplary science-fiction novel of the universe of simulation.
28. Eleanor Arnason, 'The Warlord of Saturn's Moons' (1974), *The Norton Book of Science Fiction,* p. 305.
29. Ibid., p. 305.
30. Ibid., p. 305.
31. Ibid., p. 309.
32. Ibid., p. 312.
33. Jacob Kessel, 'Invaders' (1990), *The Norton Book of Science Fiction,* p. 848.
34. Ibid., p. 850.

35. Ibid., p. 843.
36. Ibid., p. 847.
37. This is in significant contrast to, for example, L. Sprague De Camp's classic *Lest Darkness Fall* (1941), a time-travel power fantasy *par excellence*, in which a twentieth-century hero travels to Rome in 535 AD and manages to prevent Europe's 'Dark Ages' from taking place.
38. John Clute, *Look at the Evidence*, p. 281.

Index

CPSIA information can be obtained
at www.ICGtesting.com
Printed in the USA
LVOW13s1031270217
525539LV00025B/707/P